**Photovoltaic Module
Reliability**

Photovoltaic Module Reliability

John H. Wohlgemuth
Virginia
USA

Registered Offices
John Wiley & Sons, Inc., 111 River Street, Hoboken, NJ 07030, USA
John Wiley & Sons Ltd, The Atrium, Southern Gate, Chichester, West Sussex, PO19 8SQ, UK

Editorial Office
The Atrium, Southern Gate, Chichester, West Sussex, PO19 8SQ, UK

For details of our global editorial offices, customer services, and more information about Wiley products visit us at www.wiley.com.

Wiley also publishes its books in a variety of electronic formats and by print-on-demand. Some content that appears in standard print versions of this book may not be available in other formats.

Library of Congress Cataloging-in-Publication data applied for

HB ISBN: 9781119458999

Cover Design: Wiley
Cover Image: © foxbat/Shutterstock

Set in 9.5/12.5pt STIXTwoText by SPi Global, Pondicherry, India
Printed and bound in Singapore by Markono Print Media Pte Ltd

10   9   8   7   6   5   4   3   2   1

Contents

Acknowledgments

In my 50-year PV journey from graduate student through retirement from National Renewable Energy Laboratory (NREL), I have been assisted and befriended by many colleagues along the way. I would like to take this opportunity to acknowledge some of them.

The first one I would like to acknowledge is Hillard B. Huntington of RPI, my PhD thesis advisor. Then, I would like to acknowledge my two Post-Doctorate advisors, Donald Brodie of the University of Waterloo who provided me with the opportunity to begin research work in semiconductors and Martin Wolf of the University of Pennsylvania who gave me the opportunity to begin my career in photovoltaics.

I certainly need to acknowledge the impact of Joseph Lindmayer and Peter Varadi on my career. They offered me the opportunity to work for Solarex, one of the first PV-module manufacturers. Joseph provided me with my first education in PV while Peter provided the lessons in business. From the early days of Solarex, I would like to acknowledge the assistance and training I received from Chuck Wrigley, John Goldsmith, and Ramon Dominguez. I would also like to acknowledge the efforts of those who actually made the solar cells and modules for the technology group during this time including Don Warfield, Joe Creager, Dan Whitehouse, George Kelly, and Tim Koval. It was John Corsi and Alain Ricaud who brought me back to Solarex after a brief stint out of PV. Alain is also the one I must thank for getting me involved in IEC standards where I spent 40 years as a member of Working Group 2 (Modules) of IEC TC82 – the Technical Committee on PV. From that era at Solarex, I would also like to acknowledge the support from, and collaborations with, Ray Peterson, Steve Shea, Mohan Narayanan, Jim Emming, and Jean Posbic. I would also like to acknowledge the contributions of Mark Conway and David Meakin who made most of the measurements for technology.

Solarex then became BP Solarex and finally just BP Solar. It was Peter Bihuniak who changed my focus from making more efficient and less expensive cells and modules to module reliability, performance measurements and standards. From the BP Solar days, I would like to recognize Steve Ransome, Danny Cunningham, David Carlson, Paul Garvison, and Zhiyong Xia for their contributions to PV and for their collaborations with me. I would also like to recognize the engineering staff from module technology, including Andy Nguyen, Jay Shaner, Jay Miller, Dinesh Amin, and Neil Placer.

I would like to thank Sarah Kurtz for facilitating my move to NREL. Sarah gave me the opportunity to transition my work in PV-module reliability to NREL and beyond where it could impact much of the PV world. While at NREL, I would like to acknowledge support and collaboration with Sarah, Michael Kempe, Nick Bosco, and David Miller. I would also like to recognize a number of additional members of the NREL staff for their valuable work in PV including Dirk Jordan, Tim Silverman, Mike Deceglie, Ingrid Repins, and Peter Hacke.

There are also a number of people from outside the organizations that I worked for whom I would like to acknowledge for their collaborations with me and for their contributions to PV reliability including Bob Hammond of ASU, Govindasamy TamizhMani (Mani) of ASU, Neelkanth Dhere of FSEC, and Jim Galica of STR and Stevens Urethane.

I would also like to recognize specific contributions to IEC PV standards. PV standards would not be nearly as effective or as advanced as they are today if it weren't for the efforts of the secretaries to IEC TC82, particularly Jerry Anderson, Howard Barikmo, and George Kelly. I would also like to thank Steve Chalmers and Alex Mikonowicz, long-serving technical advisors to the US TAG for IEC TC 82, for their efforts. I would particularly like to acknowledge Tony Sample, who took over the convenorship of WG2 when I retired. Chris Flueckiger and Kent Whitfield should also be recognized for the support they gave in developing the US version of the IEC module safety standard.

I would also like to acknowledge those who worked with me to found PVQAT - namely Sarah Kurtz, Tony Sample, Michio Kondo, and Masaaki Yamamichi. I would like to thank my wife, Beth for all of her support and encouragement during my four decade PV career and particularly during the writing of this book.

Finally, PV in general and PV reliability in particular, has been a great technology field to work in with many dedicated professionals working together to advance the technology, and to solve the industry's problems and eventually the world's energy problems. For any students using this book for their University course work, I applaud your choice of a technology to study and hope that this book advances your knowledge in PV reliability and maybe encourages you to consider it for your career choice.

1

Introduction

Photovoltaics (PVs) is the direct conversion of light into electricity. Typically, this means generation of electricity from sunlight, a renewable energy process without release of pollution or greenhouse gases. PV is one of the renewable energy sources that offers the potential to replace burning of fossil fuels and, therefore, to slow the growing effects of global climate change.

When the author began working in PV at Solarex in 1976, the entire worldwide annual production of PV modules was less than 100 kW. Numerous groups [1, 2] are predicting that more than 100 GW of PV modules will be produced and shipped in 2019. That is a growth in production volume of more than six orders of magnitude across a span of little more than 40 years. PV has gone from a small niche business, providing electricity for remote power applications to a mainstream, electric power producing industry. It has been estimated that PV provided approximately 3% of the world's electricity in 2018. PV and wind have been the two fastest growing sectors in commercial electricity production in the world for a number of years [3].

So why has PV been so successful? Certainly, the fact that PV is a clean, non-polluting source of electricity is very important. The fact that the prices for PV modules have fallen dramatically from more than $40 per peak watt in the 1970s to around $0.40 per peak watt today [4] has also certainly helped. Today's price level makes PV one of the lowest-cost sources of electricity in the world. However, none of these would really matter if PV technology, in general and PV modules in particular, were not very reliable and have long service lives. How many products do you know of that have to work outdoors in all kinds of weather and yet are provided with a 25-year warranty? Since PV is a solid-state process, there are no moving parts and very little to wear out so PV modules should be able to operate for a long time. Longevity is critical to the value of PV since the investment to install a PV system is made in the beginning and then the income accumulates over years as the electricity is sold. Without reliable PV products, no one would be risking billions of dollars to purchase and install the PV power systems that have made PV the success it is today.

This book is about Photovoltaic Module Reliability. Modules are usually the most reliable component within a PV system, but it is important to continue to study their reliability. Modules are the most expensive and by far the hardest to replace component in the PV system if something goes wrong. Any large-scale failure of modules could result in dra-

Photovoltaic Module Reliability, First Edition. John H. Wohlgemuth.
© 2020 John Wiley & Sons Ltd. Published 2020 by John Wiley & Sons Ltd.

matic reductions in growth of the PV power industry. This book has been written from a historical prospective to help the reader understand how PV modules were able to achieve 25-year lifetimes and a well-deserved reputation for reliability. Knowing how this level of reliability was achieved, can help in understanding how to maintain similar reliability in the future with new generations of products.

This introductory Chapter will provide some background for the in-depth look into module reliability in subsequent chapters. The first section provides a brief history of PV. The second section discusses some of the different types of materials and devices used for commercial solar cells. The third section covers module packaging, including their purpose; the types of structures used for different modules, and a brief introduction to the types of materials used in today's commercial PV modules. The next section discusses what the author means by the reliability of PV modules as well as introducing several other terms that will be used throughout the book. The final section in this first chapter provides a brief overview of what is contained in each of the subsequent chapters.

1.1 Brief History of PVs

To outsiders, it may seem like PV appeared quickly out of nowhere. In reality, the technology has been under development for a long time. Let's take a brief look at the history of PV.

- Edmond Becquerel discovered the PV effect in 1839. So, PV is certainly not a new technology [5].
- Albert Einstein published a paper explaining how the PV effect worked in 1905. In 1921, he received the Nobel Prize in Physics for his discovery of how the PV effect works [6]. PV was a lot less controversial than relativity at that time.
- In 1954, a group at Bell Laboratories developed the first practical silicon solar cells [7].
- The Bell Laboratories development was just in time for PV to provide power for all of the US satellites designed to perform in space for more than a few days. Vanguard 1 launched in March 1958, was powered by PV and continued to transmit data back to earth for six years [8], while purely battery-powered satellites typically only provided data for a few months. Most US satellites continue to use PV as their primary energy supply. This means that your satellite weather forecasts, long-distance communications and TV signals have always been powered by PV. Most of us have been taking advantage of PV in this way for decades.
- In the 1970s, a terrestrial PV business was developed by two small companies (Solarex Corporation and Sensor Technology) to provide power systems for remote applications. In these remote site applications, PV was cost effective even at a $20–$40/Wp cost for modules. These remote applications included telecommunications, weather stations, navigational aids, water pumping, fence charging, remote vacation homes and cathodic protection.
- Partially because of the US energy crises of 1973 and 1979, the US government began an expanded effort in renewable energies. The Flat-Plate Solar Array (FSA) Project, funded by the US Government and managed by the Jet Propulsion Laboratory, was formed in 1975 to develop the flat-plate module and array technologies needed to attain widespread

terrestrial use of PV [9]. While many important developments came out of this effort, three had particularly important impacts on future PV efforts. The first was the initial efforts to evaluate the potential for PV technologies to undergo significant cost reductions and, therefore, eventually compete with traditional sources of electricity. The second was the proof that PV modules could be assembled into larger-scale systems that could then power real-world applications. The third was the Jet Propulsion Laboratory (JPL) Block Buy Program to be discussed in more detail in Chapter 4. This effort led to the use of acceleration stress tests and the establishment of qualification tests for PV modules. This went a long way toward improving the reliability and increasing the service life of PV modules.

- The election of Ronald Reagan to the Presidency in 1980 resulted in a huge setback for PV. The national budget for PV was slashed drastically and PV research in the US dropped to a small fraction of what it had been under Jimmy Carter's Administration. As a result, PV progress slowed appreciably and the center of PV development shifted away from the US to Europe and Asia.

- In 1994, The Japanese Ministry of Economy, Trade and Industry (METI, formerly called MITI) launched a subsidy program for residential PV systems with an overall goal of installing 4.82 GW of PV by 2010. The program was launched with a subsidy of 50% of the cost of the PV system. The program attracted homeowners not only because of the subsidy, but also because the residential electricity rates in Japan were about 24 Yen/kWh equivalent to about $0.24/kWh at that time, among the highest residential rates in the world. On the other hand, mortgage interest rates in Japan are low (1–2%) and were extendable to cover the costs of residential PV systems. From 1994 until 2003, the Japanese PV market grew steadily by about 30% a year even though the subsidy level was reduced every year. By 2003, the Japanese market was the largest in the world, representing >40% of the world's PV demand.

- Germany used a different approach to subsidizing PV systems. Rather than assisting with the initial purchase as was done in Japan, the German program pays a rate or tariff-based incentive on the electricity actually produced. The PV system owner is paid a specified rate for each kWh produced by the PV system. While such a program had been ongoing in Germany since the late 1990s, a change in the structure of the incentive program that began in 2004, resulted in explosive growth of the German PV market in 2004 and 2005. The feed in tariff rates were established at €45.7 ¢/kWh for ground mounted systems (with a 6.5% annual reduction), at €57.4 ¢/kWh for rooftop residential and at ~€54 ¢/kWh for rooftop commercial (with a 5% annual reduction). In addition to the feed in tariff, preferential loans were made available for PV. This whole program was focused on PV's CO_2 lowering potential. In Germany, the incentive for PV went from environmental to economic with significant financial returns earned via investment in PV systems. In 2004, the German PV market more than doubled, overtaking Japan as the world's largest PV market. The dramatic market growth was actually constrained in late 2004 and 2005 by lack of module availability.

- In the mid-2000s, China began to realize that PV was going to be an important energy industry and so Chinese companies began manufacturing PV modules. By 2010, an appreciable share of PV modules in the world was being produced in China. By 2015/2016 a majority of all PV modules manufactured in the world were being made in China. The volume has continued to grow since then. The combination of large-volume production and low costs for labor and infrastructure has led to dramatic decreases in the selling price for PV modules. According to Paula Mints [4] average worldwide module costs have gone from $3.00 to $4.00 per watt in 2007/2008 to less than $0.5 per watt in 2018.

The combination of market need, driven by feed in tariffs and the building of large factories in China led to an explosive growth in the PV industry. With the low cost of modules today, PV is competing successfully with conventional forms of electricity generation. Large (>100 MW) PV systems are being installed around the world with China leading the way both as producer and consumer of PV modules. For a detailed account of the history of how the Photovoltaic Industry began see Peter Varadi's book entitled "Sun Above the Horizon" [10]. For a detailed description of how the PV industry was able to grow so rapidly see Peter's book entitled "Sun Towards High Noon" [11].

1.2 Types of PV Cells

Solar cells are the devices that convert sunlight into electricity. Solar cells are made of semiconductor materials and use a junction to separate the carriers of different charges, that is, to separate the electrons from the holes to create a voltage. There are a number of good articles on how solar cells work including one by the author [12] as well as others in print [13] and on line [14]. This book will not go into great detail about how solar cells work, but will discuss them in terms of how they impact the reliability of PV modules.

A variety of materials have been used commercially to make solar cells. The list below discusses those that have been successfully utilized in terrestrial PV modules and are available commercially today.

Crystalline silicon materials: Crystalline silicon materials were the first utilized for fabrication of solar cells by Bell Laboratories [7]. They were also the first used for space applications and then the first commercialized for remote power terrestrial applications. Crystalline silicon technology has been, and still is, the dominant commercial PV material, with more than 90% of the terrestrial PV market in 2018 [4].

Crystalline Si is a wafer-based technology. Today, all of the commercial wafers are cut from ingots. There are two different types of crystalline silicon materials, single or mono-crystalline and multi-crystalline. They differ in the way the silicon ingots are grown. The single-crystal ingots are grown by a crystal growth process called Czochralski (CZ), where a single-crystal seed is dipped into a bath of molten silicon. As the seed is withdrawn, the silicon grows the same orientation as the seed. This is the same process that is used to make the wafers for many semiconductor devices. Single-crystal ingots are round, but round wafers do not pack very well into rectangular modules so, in most cases, the round wafers are cut into square or pseudo-square wafers for PV. Multi-crystalline Si is grown using direction solidification in a crucible or mold. This process was developed specifically for PV in the late 1970s [15]. According to Paula Mints [4], in 2018 multi-crystalline Si technology accounted for slightly more than 50% of the worldwide shipments of PV modules, while mono-crystalline Si accounted for approximately 45%.

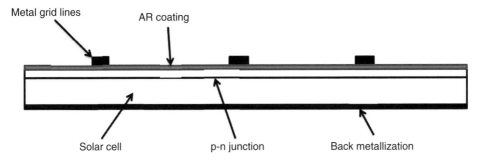

Metal grid lines AR coating

Solar cell p-n junction Back metallization

Figure 1.1 Cross-sectional drawing of screen-printed cry-Si solar cell.

For both mono- and multi-crystalline Si, the ingots are cut into wafers using wire saws. The resultant wafers are then processed into cells and the cells incorporated into modules in very similar ways. There are a number of cell processes in use today. Let's take a look at screen-print technology (one of the simpler methods) since most of the cell components will be similar no matter which cell design is used. Figure 1.1 shows a diagram of the important features of a screen-printed cry-Si solar cell.

The solar cell is fabricated on the Si wafer. A p-n junction is grown in the front (sun side) usually by diffusion. Metal grid lines are printed on the front to collect the current generated by the cell. The back side is also metallized. The drawing shows continuous back metallization though a grid can also be used on the back. Finally, because bare silicon is quite reflective, an antireflective coating (AR) is usually applied to the front. For monocrystalline Si the front surface is usually textured to increase optical absorption. This is one of the reasons that mono-crystalline cells are usually a few percent higher in efficiency than multi-crystalline cells. It is important to remember the different components in the cell when addressing reliability, because it is usually not the Si wafer or the p-n junction that degrade, but rather the contacts, the interconnections between cells or even the AR coating.

Screen-printed cry-Si cells are typically fabricated on 15.6× 15.6 cm wafers. Over the years, the wafer thickness has been slowly reduced to lower the cost of the Si used in the module. Today, Si wafers are typically less than 200 μm thick, which makes them more susceptible to breakage than the thicker wafers that were used in the earlier days of PV. Today most commercial screen-printed cells have a conversion efficiency in the range of 15–19% with the multi-crystalline Si cells at the low end and monocrystalline Si cells at the high end.

A number of commercial cell manufacturers have utilized specialty structures in order to increase cell performance. Several examples are:

- SunPower has commercialized cells with all of the contacts on the back of the cell. This cell structure eliminates front surface shadowing and can provide improved collection of the high currents associated with large area cells. This cell structure requires silicon substrates with very high lifetimes and excellent front surface passivation. SunPower offers their commercial cells with conversion efficiency up to 22.7% [16].
- The HIT (Heterojunction with intrinsic thin layer) cell uses heterojunctions between a-Si and crystalline silicon to produce a much higher voltage than the standard p-n junction.

So HIT cells have high efficiency with research cells reported at 25.6% [17] and commercial modules available from Panasonic with reported efficiencies of 19.7% [18].

- One of the latest methods for increasing cell efficiency is the use of PERC Technology (Passivated emitter rear cell). PERC cells use the same screen-printed front surface as standard screen-print technology, but the rear is modified by replacing the full metal coverage with a passivated dielectric layer with small area back contacts. PERC improves the cells by reducing the back-surface recombination of carriers and improving the reflection of long wavelength light back into the cells [19]. Solar World has reported PERC cell efficiencies of 22% [20]. There have been a number of forecasts that PERC will gain significant market share in the next few years.

Each solar cell produces a voltage determined by the semiconductor junction. For crystalline Si the typical p-n junction cell has an open circuit voltage of 0.6–0.7 V and a peak power voltage of around 0.5 V. Therefore, to reach useful voltages a number of cells are connected together in series into a module. Each cell has metal contacts typically copper ribbons attached to the front metal grid and then to the back metallization of the next cell in the string. Ribbons from the front of the cell before it in the string are attached to its back metallization. In this way cell voltages are combined to reach useful levels. Today, most cry-Si power modules have 60 or 72 cells in series. The next section will talk about the module packaging in more detail.

Thin Films Materials: In the thin-film approach, the various layers of a PV device are deposited directly onto a substrate or superstrate. The advantages of thin films are the potential ability to:

- Dramatically reduce the amount of material utilized. In some cases, the active semiconductor only needs to be a fraction of a micron thick in order to absorb most of the incident sunlight.
- Directly integrate into a higher voltage module, thereby eliminating much of the handling and labor necessary to produce cry-Si modules.

For monolithically integrated thin film modules, the deposition processes and the integration processes are described in Table 1.1 and the resultant structure shown in cross section in Figure 1.2. Figure 1.2 is not drawn to scale as the thin film layers and the three different scribe lines are exaggerated in size so you can see them. In thin film devices, the semiconductors have considerably higher resistance than cry-Si so metal grids do not work well. Instead, a Transparent Conductive Oxide (TCO) is often used on the semiconductor surface to provide a conductive path as well as serving as the AR. Because the TCO layers are not as conductive as a metal grid, thin film solar cells are usually long and skinny, so that the collected current only flows a short distance across the cell to the next cell. Monolithically integrated thin film modules often have a large number of skinny cells connected in series, resulting in higher voltage, lower current and reduced series resistance losses.

Thin film layers of materials are more susceptible to corrosion than bulk materials. When thin film layers are deposited directly in contact with glass, they may also be susceptible to ion flow in the glass [21]. This is why many thin film modules are protected in packages that are designed to keep moisture out for the lifetime of the product.

Table 1.1 Process for making monolithically integrated thin film modules.

Step #	Step Name	Description
1	Transparent Conductive Oxide TCO	Deposit TCO layer onto the substrate
2	P1 Scribe	Scribe to remove TCO from selected areas
3	Semiconductor Deposition	Deposit the semiconductor layers that make the solar cells
4	P2 Scribe	Scribe to remove the Semiconductors from selected areas
5	Metallization	Deposit the back metallization
6	P3 Scribe	Scribe to remove the Metallization from selected areas

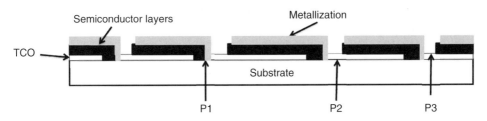

Figure 1.2 Cross-sectional drawing of monolithically integrated thin film module.

According to Mints [4], thin films only represented about 5% of total worldwide module shipments in 2018, a declining percentage from previous years. The thin-film PV commercial market includes the following three materials.

1) Cadmium telluride (CdTe): CdTe is a well-known semiconductor often used in high performance infrared (IR) sensors. CdTe absorbs visible light very strongly, so very thin films (1–2 μm) are sufficient to absorb most of the sunlight. Commercial CdTe modules are typically fabricated on glass with a structure similar to the cross-sectional drawing in Figure 1.2. First Solar is by far the largest supplier of CdTe modules. First Solar now offers large area, monolithically integrated CdTe modules with efficiencies of up to 18% [22]. CdTe technology is especially attractive because of low-cost manufacturing. In 2018, First Solar, selling just CdTe modules shipped about 3% of the world's PV modules [4].

2) Copper Indium Diselenide (CIS) and Copper Indium Gallium Diselenide (CIGS): Both $CuInSe_2$ (CIS) and $Cu(InGa)Se_2$ (CIGS) are ideal PV-absorber materials. The band gap is near the optimum for absorbing the terrestrial spectrum. These materials have strong optical absorption so very thin films (~1 μm) are sufficient to absorb most of the sunlight. Grain boundaries and surfaces in CIS and CIGS are electronically benign, so simple polycrystalline films yield reasonably high efficiency PV devices without complex grain boundary passivation. Typical CIS and CIGS cells use thin CdS layers to form p-n junctions, molybdenum for ohmic contacts to the CIS or CIGS, and transparent conductors like zinc oxide and indium-tin oxide for contact to the CdS and to serve as an AR

coating. In most cases, CIS and CIGS cells are deposited upside down, meaning that the back metal is deposited first and the cells are deposited from back to front onto the metal. So, CIS and CIGS modules often have the cells deposited onto the back glass of a glass/glass package or onto a metallic substrate that is then diced up to produce individual cells that are packaged like you would wafer-based cells. While CIS and CIGS are attractive PV materials, it has taken a long time to move from the laboratory to commercialization. Two reasons for this appear to be the difficultly in scaling to larger sizes and larger volumes, related to uniformity of the deposited films, and sensitivity to environmental stresses. Reliability issues relate to the humidity sensitivity of the contacts and TCO necessitating hermetic sealing to achieve long-term life. Solar Frontier from Japan is the world's largest manufacturer of CIS PV modules offering large area, monolithically integrated CIS modules with conversion efficiencies up to 15% [23]. In 2018, CIS/CIGS made up about 1% of the worldwide shipments of PV modules [4].

3) Amorphous Silicon (a-Si): Alloys of a-Si made of thin-film hydrogenated silicon (a-Si:H) can be deposited on either glass superstrates or flexible metallic substrates. Because of the low minority carrier lifetimes in doped a-Si, p-i-n (p-type, intrinsic, n-type) cell structures are utilized rather than normal p-n junctions. Even in undoped a-Si, the minority carrier lifetimes are short and because of the Staebler-Wronski effect, in which light induces additional electrically active defects, the thickness of individual layers must be minimized. Therefore, to achieve reasonable efficiencies, multi-junctions are usually utilized. Commercial a-Si is typically fabricated with a structure similar to the cross-sectional drawing in Figure 1.2. The maximum reported efficiency for a-Si modules is about 12% [17], which is less than the efficiencies of commercially available CdTe or CIGs modules. Commercial a-Si modules have efficiencies in the 5–8% range after light stabilization. Most a-Si products are used in consumer-type products (watches, calculators, etc.) not for power production.

Over the years, there have been a number of efforts to use optical concentration to increase the output power that can be obtained from PV devices. Some of these efforts have resulted in laboratory systems with high efficiencies and good performance. However, to date, none have been able to establish a sustainable commercial market. For this reason, reliability of concentrator PV will not be discussed in this book.

1.3 Module Packaging – Purpose and Types

While the solar cells actually produce the electricity, the module package is important for the continued operation of the solar cells. Often the costs associated with the packaging exceed the costs of the cells themselves. Historically, it is usually the package that fails first, ultimately leading to degradation of the cells, conductors, connectors and diodes resulting in failed or degraded modules. The PV module package provides for the following functions:

- Mechanical support – holding the cells in place pointing toward the sun.
- Dielectric protection – keeping the high voltage away from people and keeping current from flowing out of the array circuit (to ground or in a loop) where it has the potential to cause a fire.

Table 1.2 Typical commercial module constructions.

Glass Superstrate: Cry-Si Cells

- Glass/encapsulant/cry-Si cells/encapsulant/backsheet
- Glass/encapsulant/cry-Si cells/encapsulant/glass

Glass Superstrate: Thin Film Cells

- Glass/thin film cells on front glass/encapsulant/glass with edge seal
- Glass/thin film cells on front glass/encapsulant/backsheet
- Glass/encapsulant/thin film cells on back glass/substrate

Flexible substrates

- Transparent frontsheet/encapsulant/thin film cells/flexible substrate
- Transparent frontsheet/encapsulant/cry-Si cells/encapsulant/flexible substrate

- Protection of the cells, diodes and interconnects from the weather (UV, rain, humidity, hail, etc.)
- Coupling of the maximum amount of light energy possible into the solar cells (at all angles at the wavelengths that the cells can utilize).
- Cooling of the cells to minimize their temperature increase.

There are really just a few types of module constructions that make up the vast majority of commercial PV modules. Table 1.2 provides a list of the types of typical commercial module constructions. A vast majority of PV modules use glass as the front surface because of its excellent optical properties and as we will see in Chapter 2 as the main structural support because of the low thermal expansion coefficient of glass.

The first construction for cry-Si modules (Glass/encapsulant/c-Si cells/encapsulant/ backsheet) has certainly been used on more modules than any other and still remains the most popular in the industry. Figure 1.3a shows a cross-sectional drawing of a Glass/ encapsulant/cry-Si cells/encapsulant/backsheet module construction. The second construction for c-Si modules substitutes a second glass layer for the standard backsheet as shown in cross-section in Figure 1.3b. This type of construction is becoming more popular especially for use in bifacial designs (modules that produce electricity from light that falls on both sides, not just the front). The third construction for c-Si modules is a flexible design. This is shown in Figure 1.3c. Flexible modules are usually designed as portable power supplies to be carried and deployed when needed. They are not designed for continuous outdoor exposure.

Thin film cells are deposited onto a foreign substrate. These substrates can be glass where the cells are deposited right side up or upside down depending on the technology of the

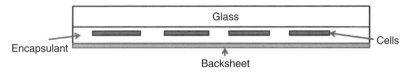

Figure 1.3a Cross-sectional drawing of glass/encapsulant/cry-Si cells/encapsulant/backsheet module.

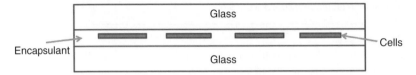

Figure 1.3b   Cross-sectional drawing of glass/encapsulant/cry-Si cells/encapsulant/glass module.

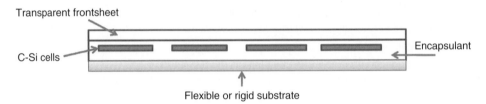

Figure 1.3c   Cross-sectional drawing of flexible cry-Si module.

particular thin film material being used. Figures 1.4a and 1.4b show the cross section of these two types of module constructions. In Figure 1.4a, the thin film is deposited on the backside of the front glass. This is typical of how CdTe and a-Si modules are fabricated. Figure 1.4a has been drawn with edge seals as this is typically how CdTe modules are fabricated today. The edge seals are designed to keep moisture from reaching the active cell area for the lifetime of the product (typically warrantied by the manufacturer for 25 years). In Figure 1.4b, the thin film is deposited on the front side of the back glass. This is typical of how CIS and CIGS modules are fabricated. Figure 1.4b has also been drawn with edge seals, but edge seals are not as prevalent in these types of modules. In this case, the superstrate can also be made of glass though other materials are often used.

In some cases, thin films are deposited in large areas and then cut to cell size afterwards. Basically, creating wafers out of thin films which then have to be electrically connected in series just like cry-Si cells. Figure 1.4c shows the typical construction used for such thin film modules, although any of the constructions used for cry-Si wafers could also be used

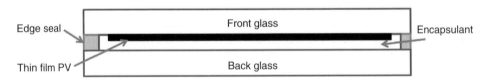

Figure 1.4a   Cross-sectional drawing of front glass/thin film cells/encapsulant/back glass modules.

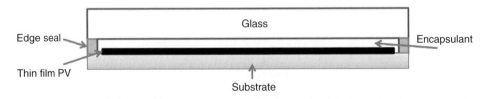

Figure 1.4b   Cross-sectional drawing of front glass/encapsulant/thin film cells/substrate modules.

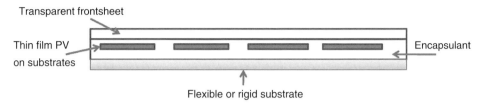

Figure 1.4c Cross-sectional drawing of module structures for thin film wafer like cells.

to package these thin film "wafers." Some CIGS modules have been made with these types of cells. They are particularly of interest for fabrication of large-area flexible modules.

There are only a handful of materials that appear in these drawing so let's briefly take a look at the properties required and those typically selected for use in PV modules.

Glass: When glass is used as the superstrate, one of the properties of interest is the optical transmittance over the wavelength range, that solar cells can effectively use the photons, from about 300 nm to 1100 nm for cry-Si for example. To maximize performance without significantly increasing the cost, most cry-Si modules and some thin-film modules are built using low iron glass which has better transmittance than the standard soda lime (window) glass. Some thin-film PV modules do use regular soda lime glass to keep the cost down. In addition, most cry-Si modules use tempered or heat-strengthened glass to provide added strength to withstand wind and snow loads as well as hail impact. Some thin-film modules can't use heat-strengthened glass because the thin-film deposition process occurs at such a high temperature that the heat strengthening would be removed from the glass. In this case, the modules are usually built with double glass (glass on front and back) to provide the strength necessary to survive in the field.

Encapsulant: The encapsulant is the material that surrounds the cells and the "glue" that holds the whole package together. The encapsulant should provide good adhesion to all of the other components within the module so that everything in the package stays stuck together for 25 or 30 years. This is usually assisted by addition of a primer into the encapsulant formulation itself. Of course, any of the encapsulant material that is used in front of active solar cells must be optically transparent and resistant to UV exposure. Since the encapsulant surrounds the solar cells, it helps to provide electrical isolation. So, materials used as encapsulants must have low-bulk conductivity (or high-bulk resistivity) to minimize flow of leakage currents. Some encapsulants are cross-linked during module lamination to provide stability at the high temperatures at which they operate. Others, however, do not have to be cross-linked since they are stable enough not to flow or creep at typical module operating temperatures. Typical examples of materials used as encapsulants are listed below:

- Ethylene vinyl acetate (EVA) has been used in more modules than any other encapsulant as it is reasonably priced and readily available as a formulated film for PV with primers, cross-link agents and UV stabilizers incorporated into the film itself.
- Silicones were used in the early days of PV and worked well in the field but were abandoned due to their high costs and because liquid encapsulants were more difficult to use in manufacturing.
- Polyolefins are similar to EVA and are also available as formulated sheets. They have become more popular as a replacement for EVA in recent years.

- Ionomers were used in the past, especially by Mobil Solar and ASE Americas. They were typically used in glass/glass constructions but the particular formulation used had problems with delamination in the field probably due to less than ideal adhesion to the glass.

<u>Backsheets:</u> As the name implies, the backsheet is the outside material on the back or non-sun side of the PV module. The functions of the backsheet include:

- Protecting the rest of the module from the weather – rain, snow, hail, etc.
- Screening the materials inside it from UV.
- Providing protection from the high voltage within the module, so backsheets must have high resistance and high dielectric strength.
- Providing protection from accidental exposure to the active components within the module. So, the backsheet must have high-tensile strength and be scratch resistant.
- The backsheet is the first line of defense in case the module catches fire so its properties will impact the fire rating that a module can obtain.
- The backsheet must provide for secure bonding of junction boxes, connectors, frames and/or mounting rails.

Backsheets are usually comprised of multi-layers of material as the different layers provide different functions. Often one of the layers (usually the center layer) is a poly(ethylene terephthalate) (PET) or polyester to provide the dielectric isolation and high resistance. The outer layer has to provide UV resistance to the layers inside and is most often composed of a fluoropolymer. The inner most layer must bond well to the encapsulant. One typical example of a multilayer backsheet is Tedlar/polyester/Tedlar.

<u>Edge Seals:</u> Modules that are constructed with impermeable (or extremely low permeability) front and backsheets designed to protect moisture-sensitive PV materials, may suffer from moisture ingress from the sides. Edge Seal materials are low-diffusivity materials that are placed around the edges of a module between the impermeable front and backsheets to prevent moisture ingress. Edge seals were borrowed from the insulated glass industry where they are used to keep moisture from penetrating between the two panes of glass. In addition to restricting moisture ingress, edge seal materials must have high electrical resistivity to provide electrical insulation as frontsheets and backsheets do. To continue to perform these functions for the lifetime of the module, edge seal materials must remain well adhered to the front and back sheets of glass. Edge seal materials are usually made of Polyisobutylene and filled with desiccants to keep moisture from penetrating throughout the useful life of the module.

<u>Frontsheets:</u> Frontsheets must meet all of the same requirements as backsheets, with the additional requirement of having high optical transmittance over the wavelength range that solar cells are effective, for example from about 300 nm to 1100 nm for cry-Si. Glass is used the most as a frontsheet, but some modules are made using fluoropolymer frontsheets, particularly Tefzel or Ethylene tetrafluoroethylene (ETFE).

1.4 What Does Reliability Mean for PV Modules?

Since the book is entitled "Reliability of Photovoltaic Modules," it is important to explain what reliability means in this context. In the simplest sense, reliability of a product means that it performs the promised functions for the expected lifetime. For a PV module, the promised function is continued production of the specified amount of electricity in a safe manner. Typically, the product warranty will specify both the amount of power retained (something like retention of 80% of initial power) and the time period over which it should retain this power level (often 25 years). So, if a particular module fails to produce this amount of power or if it becomes unsafe (maybe it has too high a leakage current or it falls out of the frame), then it is considered a failure. For a particular type of module, the term reliable would indicate that a certain percentage (maybe something like >95%) are still operating within the specified power limit throughout the length of the warranty period.

Another word often used to describe PV modules is their durability. The dictionary defines durability as "the ability to withstand wear, pressure, or damage." PV modules must be durable since they are exposed to the stresses of the outdoor environment when deployed. For modules, durability has come to mean the continued ability to provide output power over its lifetime. Rather than focusing on failure as reliability does, durability focuses on maintaining the output power level. As a module ages, it may slowly degrade in output power. This is usually reported as an annual degradation rate although the rate is often not linear, but that is the way it is reported for simplicity. So, what does it really mean if one particular module type deployed at a specific geographic location has a reported annual degradation rate of 1% after having been deployed at the site for five years? This really means that the average module output power has dropped by 5% over the five-year exposure time. The annual degradation rate is an important factor used in evaluating potential financial payback from the investment in the PV array. Investors want to know what annual degradation rate they can expect from the module type they are purchasing.

The final term to discuss in this section is module lifetime. This is a measure of how long the module will continue to produce power at the specified level in a safe manner. The warranty typically provides an estimated lifetime which should be the minimum time over which the product will continue to operate and meet the warrantied power. Once again, these values are typically 80% output power after 25 years. The real question is how the customer defines the lifetime. When does the owner of the PV modules decide that they have degraded too much and need to be retired and perhaps replaced? Experience indicates that the 80% level may be a reasonable benchmark to use because once modules have degraded below this power level, they often start having other problems like ground faults, breakage, etc. So, while there is certainly no consensus on defining module lifetime, this book will consider the useful life of a module over when the power has decreased to less than 80% of the original specification. For an array of modules, this would typically mean that the power of the array or the average power of the modules has dropped to 80% of the initial specified output power.

1.5 Preview of the Book

This book has been written from a historical perspective to guide the reader through how the PV industry learned what the failure and degradation modes were, how accelerated tests were developed to cause the same failures and degradation in the laboratory and then how these tests were used as tools to guide the design and fabrication of more reliable and longer-life modules. The following sections will provide a brief description of what will be contained within each of the subsequent chapters.

Chapter 2 begins the story with a survey of PV module failure and degradation modes that have been observed in the field. Pictures are provided for many of these to show what the failures look like and to provide the reader with a guide for their own module inspections. The book starts with failures not because PV modules are unreliable, but because identifying likely failure modes is the first step in avoiding them. For some of the failure modes, there is description of the relatively simple fixes that were used to solve some of the early problems with PV modules and therefore led to more reliable products.

The subject of Chapter 3 is accelerated stress tests (ASTs), describing what they are and how they have been used in PV to improve module reliability and lifetime. In developing ASTs, we must cause degradation. The degradation occurring in the AST must be due to the same failure mechanism we saw outdoors. The work to develop the appropriate AST for the different module failure modes is based on more than 35 years of experience. The Chapter concludes with a discussion of various ASTs that have been found to be useful for testing PV modules.

Chapter 4 introduces the concept of Qualification Tests, which are a set of well-defined ASTs developed out of a reliability program. The purpose of qualification testing is to rapidly detect the presence of known failure or degradation modes that may occur in the intended operating environment. In these tests, the stress levels and durations are limited so the tests can be completed within a reasonable amount of time and cost. The chapter starts with a history of the development of Qualification Testing for PV. It then provides a summary of the testing performed in IEC 61215; the main qualification test standard used for PV modules. The chapter then discusses how Qualification Tests have been critical to improving the reliability and durability of PV modules as well as some of the limitations of the Qualification Tests. The chapter concludes with a discussion of module safety testing including a summary of IEC 61730, the main safety test standard used for PV modules.

Chapter 5 discusses some of the tools used to better understand what has gone wrong within a failed or degraded module. So, it presents characterization tools that look to define what properties of the module (or cells) have degraded and what may have been the cause of such degradation. Methods presented include, how to analyze the I-V parameters, measurement of performance at different irradiances, visual inspection, Infrared (IR) Inspection, Electroluminescence (EL) Inspection and evaluation of adhesion.

Chapter 6 is about the use of Quality Management Systems in the manufacture of PV modules. The premise is that to continually build quality modules, the manufacturers should be using a Quality Management System that have been developed specifically for PV module manufacturing. The chapter provides the history behind how Quality Management Systems evolved in PV, indicating how successful this has been but also identifying some of the issues with a "do it yourself" system and the need for further improvements that led to the creation of the International PV Quality Assurance Task Force (PVQAT) discussed in the Chapter 7.

Chapter 7 tells the story of PVQAT including its creation and the establishment of its research goals. The three goals include development of improved accelerated stress testing for PV modules, establishment of a Quality Management System for PV module manufacturing and establishment of a conformity assessment system for PV power plants. The chapter concludes with a summary of the objectives and activities of each of the PVQAT Task groups.

Chapter 8 introduces the concept of Conformity Assessment for PV. The first part of the chapter discusses development of conformity assessment systems for PV products, mostly PV modules though some PV products like PV lanterns have also been covered. The second part discusses the extension of conformity assessment from products to PV systems deployed in the field and how this required the creation of a whole new Conformity Assessment organization at IEC called the IEC System for Certification to Standards Relating to Equipment for Use in Renewable Energy Applications or IECRE for short.

Chapter 9 discusses how to predict service life of PV modules. The chapter starts out by addressing how to determine the acceleration factors for the different ASTs that are typically performed on PV modules. It then discusses the impact of module design and control of the manufacturing process on module failure rates and how that impacts lifetime predictions. The third section explains the impact of the weather at the geographic location where the module is deployed and the type of mounting system used on module degradation and failure rates and how those impact lifetime predictions. The fourth section talks about the efforts to get the PV community to agree on one set of extended ASTs that evaluate modules for wear out. The final section discusses development of a methodology for how a PV module manufacturer could set up a system to predict the lifetime of one of their products.

Chapter 10 shifts the focus to the future. Since PV is a dynamic industry, the technology and the testing standards are constantly evolving. The first section provides an update on the changes already in progress for some of the more important module qualification and safety standards. The second section takes a longer-range view, discussing how PV module reliability is likely to change in the future and what sort of accelerated stress testing will be necessary to validate the quality of the huge volume of modules that will be produced. The book will end with a brief summary of the status of PV module reliability today.

References

1 Beetz, B. (2018). https://www.pv-magazine.com/2018/12/31/14-pv-trends-for-2019 (Accessed 28 August 2019).
2 McKerracher, C. (2019). https://www.bloomberg.com/professional/blog/transition-energy-transport-predictions-2019 (Accessed 28 August 2019).
3 IEA (2019). https://www.iea.org/statistics/electricity (Accessed 28 August 2019).
4 Mints, P. (2019). Photovoltaic Manufacturer Capacity, Shipments, Price & Revenues 2018/2019. SPV Market Research, Report SPV-Supply7.
5 Williams, R. (1960). Becquerel photovoltaic effect in binary compounds. *The Journal of Chemical Physics* 32 (5): 1505–1514.
6 All Nobel Prizes in Physics. https://www.nobelprize.org/prizes/lists/all-nobel-prizes-in-physics (Accessed 28 August 2019).
7 Chapin, D.M., Fuller, C.S., and Pearson, G.L. (1954). A new silicon p-n junction photocell for converting solar radiation into electrical power. *Journal of Applied Physics* 25: 676.
8 National Aeronautics and Space Administration. https://nssdc.gsfc.nasa.gov/nmc/spacecraft/display.action?id=1958-002B (Accessed 28 August 2019).
9 FSA. https://www2.jpl.nasa.gov/adv_tech/photovol/summary_overview.htm (Accessed 28 August 2019).

10 Varadi, P.F. (2014). *Sun Above the Horizon*. Pan Sanford Publishing.

11 Varadi, P.F. (2017). *Sun Towards High Noon*. Pan Sanford Publishing.

12 Wohlgemuth, J. Photovoltaic cells. *Kirk-Othmer Encyclopedia of Chemical Technology* https://doi.org/10.1002/0471238961.16081520070125.a01.pub3.

13 Green, M.A. (1995). *Silicon Solar Cells*. Bridge Printery Pty Ltd.

14 ChemMatters. https://www.acs.org/content/acs/en/education/resources/highschool/chemmatters/past-issues/archive-2013-2014/how-a-solar-cell-works.html (Accessed 28 August 2019).

15 Lindmayer, J. (1976). Semi-crystalline silicon solar cells. 12th IEEE PVSC in Los Angeles, Usa (15–18 November 1976).

16 Sunpower. https://us.sunpower.com/products/solar-panels (Accessed 28 August 2019).

17 NREL. https://www.nrel.gov/pv/module-efficiency.html (Accessed 28 August 2019).

18 Panasonic. https://na.panasonic.com/us/energy-solutions/solar/solar-panels/n330-photovoltaic-module-hitr-40mm (Accessed 28 August 2019).

19 REC (2014). http://www.recgroup.com/sites/default/files/documents/whitepaper_perc.pdf (Accessed 28 August 2019).

20 Fuhs, M. and Sieg, M. (2016). http://www.pv-magazine.com/news/details/beitrag/solarworld-hits-22-perc-efficiency_100022790/#axzz4Kzjs1p29 (Accessed 28 August 2019).

21 Carlson, D. (2002). Accelerated Corrosion Testing of Tin Oxide Coated Glass. *Proceedings of NREL Thin Film Module Reliability National Team Meeting*, Colorado, USA (6–10 May 2002).

22 First Solar (2019). http://www.firstsolar.com/-/media/First-Solar/Technical-Documents/Series-6-Datasheets/Series-6-Datasheet.ashx (Accessed 28 August 2019).

23 Solar Frontier. http://www.solar-frontier.com/eng/solutions/products/index.html (Accessed 28 August 2019).

2

Module Failure Modes

Initial product reliability assessments are based on the environment in which the product is going to be exposed, the outdoor terrestrial environment in the case of photovoltaic (PV) modules. Such an analysis can provide some clues as to the level of stresses to be encountered. The earliest module manufacturers understood that the modules would be required to endure exposure to the weather (rain, hail, and snow), high temperatures, UV, humidity, and thermal cycling. However, in most cases, the stresses were underestimated. The first generation of terrestrial PV modules was not very reliable nor did the modules survive for very long in the field. However, this first generation of product served an important function in that they failed in the field (often very quickly) allowing for subsequent analysis and development of accelerated stress tests to be described in Chapter 3. It wasn't until the product designs could be tested using the accelerated stress tests that reliability was significantly improved.

The study of PV module reliability starts with the identification of field failures. Over the years, there have been many reports of module field failures. Table 2.1 provides a list of field failures observed in crystalline-Si PV modules. The vast majority of deployed PV modules have been crystalline silicon so this technology has the most history. The following sections will discuss the details of each of these failure modes, with emphasis on what causes the failure, what construction or material selections make it worse or better and how the industry eventually modified their products to improve reliability and lifetime. Each subsection will also mention whether this is a failure mode likely to be observed for thin-film modules. A final section in the chapter will address failure modes specific to thin-film modules.

2.1 Broken Interconnects

The interconnect ribbons that connect solar cells together can break due to stress caused by thermal expansion and contraction or due to repeated mechanical stress. Many of the earliest modules only had one interconnect ribbon per cell so they suffered open circuits when an interconnect broke. Redundancy of interconnect ribbons was introduced to keep modules from failing prematurely. Figure 2.1 shows an Electroluminescence (EL) picture of a module with multiple broken interconnect ribbons. (EL is a useful tool for inspecting

Table 2.1 Failure modes observed for c-Si modules.

Broken interconnects

Broken/cracked cells and snail trails

Corrosion of cells, metals and connectors

Delamination/loss of adhesion between layers

Loss of elastomeric properties of encapsulant or backsheet

Encapsulant discoloration

Solder bond failures

Broken glass

Glass corrosion

Reverse bias Hot Spots

Ground faults due to breakdown of insulation package

Junction box and module connection failures

Structural failures

Bypass Diode failures

Open circuiting leading to arcing

Potential Induced Degradation (PID)

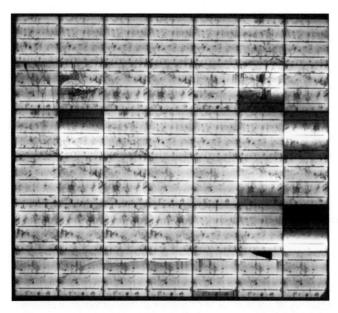

Figure 2.1 Electroluminescence (EL) picture of module with multiple broken interconnect ribbons.

Figure 2.2 Magnified picture of interconnect ribbon with a crack [1]. *Source:* reprinted from author's PVSC article.

PV modules and will be described in more detail in Chapter 5). The dark areas in the picture are the regions where broken interconnect ribbons are impeding the current flow. Since the interconnects in these regions are broken, the current is no longer efficiently collected from those regions. Those regions are not producing as much energy as they should be. This module has lost approximately 10% of its initial STC power (that is peak power measured at Standard Test Conditions defined as $1000\,W/m^2$ irradiance with the reference AM1.5 solar spectrum at a junction temperature of 25 °C.)

Figure 2.2 shows a picture of a broken interconnect ribbon. The module from which this ribbon was taken also had multiple dark areas in its EL picture. This particular type of module suffered from broken interconnect ribbons because of a design flaw (the ribbons were soldered to the cell edge on both front and back so there was not enough free interconnect ribbon). Many failures of interconnect ribbons are due to poor design and usually take a significant amount of time in the field (at least five years) to have a major impact on module performance. There are a number of factors in module design that facilitate interconnect breakage including:

• Substrates (or superstrates) with high linear thermal expansion coefficients: If the substrate or superstrate has a high linear thermal expansion coefficient, it will expand and contract more during normal operation resulting in extra stress on the interconnect ribbons. Values for the linear thermal expansion coefficient of some materials relevant to PV are given in Table 2.2 [2]. Substrates made of plastics, epoxy board and even metals have high thermal expansion coefficients which stresses the interconnect ribbons during outdoor exposure causing some of the ribbons to break. Many of the earliest module designs including some of the original Jet Propulsion Laboratory (JPL)Block I modules had fiber board substrates. As will be described in Chapter 4, these modules failed open circuited in a few years, with those deployed in desert

Table 2.2 Linear thermal expansion coefficient of selected materials relevant to PV [2].

Material	m/mK ($* 10^{-6}$)
Aluminum	21–24
Copper	16–16.7
Epoxy without fillers	45–65
Epoxy wit glass fillers	36
EVA	180
Germanium	18.4
Glass	9
Plastics	40–120
PET or polyster	59.4
Silicon	3–5
Solder (tin-lead)	25
Stainless Steel	10–17

climates failing the fastest. For almost four decades now, the vast majority of PV modules have used the glass superstrate design because glass has one of the lowest thermal expansion coefficients.

- Larger cells: Larger cells expand and contract more than smaller cells and therefore exert more stress on the interconnect ribbons.
- Thicker ribbons: To reduce the series resistance losses it is advantageous to use the thickest interconnect ribbon practical. However, thicker ribbons are stiffer, resulting in the ribbon itself breaking more easily or in it stressing the solder bonds resulting in their failure (See Section 2.7).
- Stiffer ribbon: If a ribbon is soft and deforms when stressed, it is much less likely to break in a module. Over the years, the ability to make softer ribbons has dramatically improved the reliability of PV modules in terms of ribbon breakage, solder bond failure and even cell breakage.
- Kinks in ribbon: In the early days of PV, it was considered good practice to put a stress relief loop in the interconnect ribbon. However, experience showed that all these loops did was concentrate the stress in the loop instead of along the whole length of the free ribbon. The ribbons tended to fail at the loop.
- Not enough free ribbon between solder bonds on adjacent cells: It is the free region between bonds on the two cells that must take up the changes in spacing between cells caused by the thermal cycling. Even though copper (or any other metal for that matter) has a much higher thermal expansion coefficient than Si and glass, if there are only a few millimeters of free ribbon between solder bonds on adjacent cells, there isn't enough material to provide the necessary expansion and contraction. The ribbon in Figure 2.2 came from a module where the ribbons were soldered to the ends of the bus bars on both front and back of the cells. These types of modules suffered from multiple broken ribbons and excess power loss.

The improvements made in module design and the improved properties of the materials used for interconnect ribbons has dramatically reduced the instances of interconnect breakage in modern PV modules.

2.2   Broken/Cracked Cells and Snail Trails

Crystalline-Si cells can (and will) break due to mechanical and thermal stresses. Early modules suffered open circuits due to broken cells since there was only one electrical attachment point per cell for each polarity. This led to the development of the tab across design, where tabs run the full length of the cell and are soldered in multiple places along the cell length on both sides. This design has been used for most crystalline silicon modules since the 1980s. With the tab across design, a small amount of cell breakage doesn't often lead to power loss, but a large amount of breakage can.

Electroluminescence (EL) is an excellent tool for observing cell breakage. A description of how EL works will be provided in Chapter 5. Figure 2.3 shows an EL picture of a module with extensive cell breakage. The dark lines in the picture indicate cracks in the Si. However, most of these cracks just appear as thin gray lines. These do not impact the performance of the solar cell. The dark areas in the EL picture (for example, the dark area in the fifth cell from the left in the top row) indicate regions where there is no EL signal because those parts of the cells are disconnected from the module circuit and are not providing power. Such regions not only reduce the active area of the cell itself but they also lead to cell mismatch as a cell with a significant reduction in area cannot produce as much current as a full cell. Such a cell will act as a resistive load to the current produced by the remainder of the module. The module in Figure 2.3 was down 9% in peak power. We were not sure why this module suffered from broken cells while the remainder of the modules in the system had few, if any, broken cells and still provided 100% of rated power.

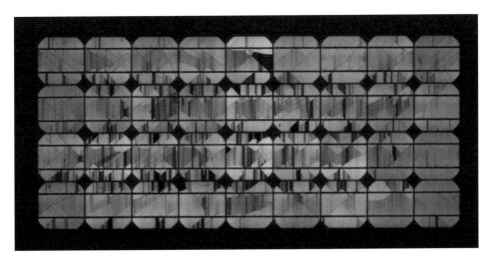

Figure 2.3   Electroluminescence (EL) picture of module with many broken cells [3]. *Source:* reprinted from author's PVSC article.

There are multiple reasons why the cells in modules crack. Some of them are related to construction of the module while others are caused by handling or improper mounting. Some of the factors to consider are:

- Cell thickness: In the 1980s, most cry-Si solar cells were 300 μm or more in thickness. Today, most cry-Si cells are less than 200 μm in thickness. The trend is for continued reduction in cell thickness to reduce the amount of Si per kW-hr of electricity produced. Thinner Si wafers break more easily than thicker ones. So we expect today's cry-Si modules to suffer from more broken cells than older modules did.
- Cell size: In the late 1970s, the terrestrial PV industry transitioned from using 2.25 in. round cells to 3 in. round cells. Today, the standard cry-Si cell is 15.6 cm by 15.6 cm or 6 in. by 6 in. (either fully-square polycrystalline Si or pseudo-square single crystal Si). The breakage rate of larger cells is higher than for smaller cells everything else being equivalent.
- Module size: In the late 1970s, modules were typically fabricated using 36 2.25 in. cells with an area of 81 square inches or 0.05 square meters. Today, standard commercial modules with 72 15.6 by 15.6 cells have an area of 3000 in.^{2} or 2 m^2. Most such modules still use 3.2 mm thick tempered glass for the superstrate. The new modules deflect more under wind and snow load. Therefore, there is significantly more stress imparted to the cells in today's larger modules.
- Crystallinity: Single-crystal wafers have cleave planes along which the cells crack easily. If the cleave plane is oriented along the bus bar the cell is more likely to crack.
- Pre-stressed or chipped cells: If the solar cells have built in stress say due to crystal growth problems or microcracks or chips caused by the processing, they will be more prone to breakage initiated at the site of the damage.
- Poor packaging of modules during shipment: Shipping can be hard on the cells within the modules. It is not unusual to observe a broken module or two at the bottom of a pallet of modules delivered to a work site. I have had personal experience shipping perfectly good modules to a test laboratory and have them arrive with multiple broken cells. Some module manufacturers test their packaging methods, but use module breakage as the pass criteria not cell breakage within the modules. IEC has now published a transportation standard, IEC 62759-1 [4] which not only stresses the modules within their transportation packaging but also stresses the modules afterwards to determine whether cells have been damaged during the transportation testing.
- Poor handling of modules during installation: The cells within modules can also be broken during installation. Everything from walking on modules to dropping modules to torqueing them down too hard onto the support structure can break cells (or even glass) or make the cells susceptible to future breakage. IEC has recently published a guideline for system installation, IEC 63049 [5] in an effort to minimize damage during installation.

Particularly because of the latter two factors, many system installers now use EL to screen the installed modules to determine whether there is significant cell breakage within their array.

Snail trails are a relatively new phenomenon that occurs as a result of broken or cracked cells. A typical example is shown in Figure 2.4. In the region around a crack in a cell, the intrusion of moisture and/or air results in corrosion of the Ag in the screen-printed metallization [6]. This can occur quickly in the field, often during the first year of

Figure 2.4 Cells with snail trails.

deployment. Early Ethylene Vinyl Acetate (EVA) modules did not exhibit snail trails, but with the introduction of large numbers of module manufacturers and material suppliers from China in the 2000s, this phenomenon began to be observed.

Fraunhofer ISE has observed that snail trail products (the corrosion on the grid lines) include silver phosphate (Ag_3PO_4), silver sulfide (Ag_2S), silver carbonate (Ag_2CO_3) and silver acetate ($Ag_2[CH_3COO]_2$) [7]. In their work, the source of the impurities was evaluated and subsequent accelerated testing was performed to determine how the corrosion might progress. Some of their main findings were:

- Silver acetate growth resulted from interaction with the EVA and was accelerated by UV exposure. There is no limit on its formation during a PV module's lifetime
- Silver sulfide results from interaction with sulfurous-containing antioxidants probably from the backsheet and was accelerated by damp heat. It is not clear whether this could be a major cause of snail trails, but it can be reduced by using a backsheet with lower oxygen transmission rate (OTR).
- Silver carbonate is formed via a reaction with metal oxides on the grid lines and it disappears after exposure to damp heat. Since there are limited metal oxides and it disappears from humidity exposure its formation is limited.
- Silver phosphate forms from a reaction with phosphorous-containing antioxidants in the EVA and does not grow under damp heat or UV exposure. Its formation is most likely limited by the amount of phosphorous available.

The finding that silver acetate is the likely culprit is consistent with the finding that different EVA's have vastly different amounts of free acetate. It would appear that those with a large amount of free acetate are much more likely to develop snail trails than EVA formulations with little free acetate.

One of the remaining open questions is whether snail trails are just cosmetic or if the corrosion of the grid metallization will ultimately lead to power loss. While some module manufacturers are telling customers it is only cosmetic, most are doing all they can to reduce the occurrence of snail trails in their products via their selection of EVA and backsheet materials.

2.3 Delamination

Delamination refers to separation of different layers within the module package. It can occur between a variety of different layers within the package. The impact on module performance and lifetime is dependent on which surface delaminates. Let's look at several different types of delamination that have been observed in the field.

- Delamination of the encapsulant from the glass: Encapsulant-glass delamination has been observed to occur in specific module types. National Renewable Energy Laboratory (NREL) has reported on such delamination in Mobil Solar and later ASE America modules at a number of sites with a variety of different climates (see Figure 2.5) [8]. These modules were fabricated with a non-EVA encapsulant which did not have as high an initial adhesion to glass as other encapsulants like EVA. This material tended to delaminate from the front glass either directly over the junction box or in the corners. This problem was observed at numerous sites often with every module exhibiting delamination directly over the junction box. However, observations at the Springerville, AZ site indicated that modules fabricated before September, 2001 are delaminating while those fabricated after September, 2001 are not delaminating. This suggests that the module manufacturer likely made a material or process change in this timeframe. The area over the junction box runs hotter than the rest of the module so it is likely that the higher temperature plays a role in the delamination. It is also possible that the mechanical stress of the junction box actually helps to pull the encapsulant off of the glass.

  Delamination between the glass and encapsulant may not indicate as severe a problem as delamination between other layers within the module. It all depends upon whether the delamination results in exposure of the cells or electric circuit to the elements. If the encapsulant stays well attached to the cells and the bus bars, it may continue to protect them from moisture. As in the case of the module shown in Figure 2.5, there is a cosmetic defect, but the module has continued to operate adequately.

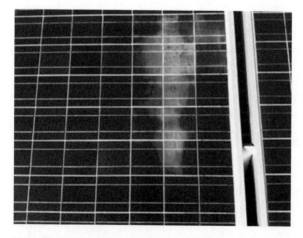

Figure 2.5 Delamination over the Junction Box of the encapsulant from front glass for ASE Module [8]. *Source:* reprinted from author's PVSC article with the permission of David Miller of NREL.

Most EVA-encapsulated modules do not exhibit this type of encapsulant-glass delamination. To help us understand why, Nick Bosco at NREL utilized a recently developed cantilever beam method to measure the adhesion between EVA and glass in old Arco Solar modules – comparing the results obtained from modules exposed in the field for 27 years with those obtained from a module continually stored in a shed for the same amount of time [9]. For the sample that had been in a shed for the whole time, the weakest interface was the EVA-cell interface with a debond energy (adhesion) of between 800 and 1000 J/m². For the sample that was exposed in the field to sunlight for 27 years, the weakest interface was the glass-EVA interface with an adhesion of between 200 and 300 J/m². When measuring the adhesion of EVA to glass in new samples, Nick typically obtains values of around 2000 J/m². So the front surface adhesion has dropped by nearly an order of magnitude. However, these modules show no evidence of delamination, indicating that 200 and 300 J/m² is adequate for module survival for more than 25 years.

In a recent set of experiments undertaken by PVQAT Task Group 5, the attachment strength between encapsulant and glass was evaluated as a function of UV exposure using various temperatures and humidity levels [10]. While these accelerated exposures did reduce the measured strength and adhesion of EVA to glass by more than 50%, they remained at levels that are still adequate to prevent delamination. This result shows that while EVA to glass adhesion will be reduced during the initial stages of field deployment, it may not continue to degrade throughout the lifetime of the module. This experiment will be described in more detail in Chapter 7.

- Delamination of the encapsulant from the cells: For EVA-based modules, the more commonly observed delamination is between the EVA and the cell surface, selectively occurring around the interconnect ribbons as shown in Figure 2.6 for a Siemens Solar module. This failure mode tends to occur in a large fraction of the modules in certain arrays.

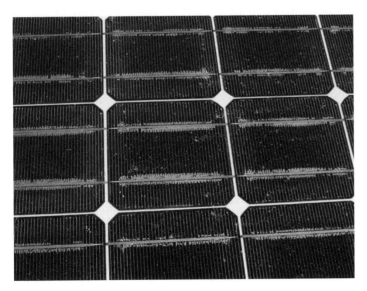

Figure 2.6 Delamination between encapsulant and cell surface in Siemens Solar Module [11]. *Source:* reprinted from author's PVSC article with the permission of David Miller of NREL.

The module in Figure 2.6 was deployed in Florida [11], where all of the modules in the array exhibited a similar level of delamination. However, there are numerous Siemens Solar arrays with little or no delamination. It is interesting to note that such delamination appears in multiple installations from Arco Solar, Siemens Solar and even Shell Solar indicating that the source of the delamination was part of the module design or manufacturing process for many years. One possibility is related to the lamination process that was used. Arco, Siemens and Shell all continued to use the standard cure EVA formulation long after most other companies had switched to fast cure EVA. Their lamination process consisted of a short cycle in the laminator followed by a much longer cure in an oven. This process appeared to work well to cure the EVA without outgassing, but it may not have been adequate to activate the silane prime resulting in occasional adhesion problems.

This type of delamination is more serious than some other types of delamination because the grid metallization is exposed and can corrode. Therefore, we expect and do see significant power loss due to increased series resistance from such modules.

This type of delamination is not observed during damp heat, humidity freeze or thermal cycling from the standard qualification testing discussed in Chapter 4. There is evidence that this phenomenon is not strictly a delamination but may be caused by generation of gas at the metallization surface. This will be discussed in more detail in Chapter 7 in conjunction with the PVQAT effort.

- Delamination of the backsheet: Delamination has also been observed between the encapsulant and the backsheet or between layers of the backsheet itself. It is often hard to distinguish between these two failure modes without destructive failure analysis. There are multiple examples where one or more layers of a backsheet have completely delaminated from the inner layer that is attached to the encapsulant. This type of backsheet delamination may lead to potential safety concerns as the backsheet ensures electrical isolation between the system voltage within the module and the outside world. However, backsheets do not play as important a role in terms of PV performance as encapsulants. We have seen PV modules in the field with their backsheets completely delaminated from the back encapsulant that are still producing power in accordance with their warranty.

2.4 Corrosion of Cell Metallization

Moisture induced corrosion of cell metallization is usually associated with significant power loss and premature failure. Corrosion observed in fielded modules is often associated with delamination of the encapsulant. Then water can condense in the void caused by the delamination. Liquid water then corrodes the cell metallization. One example is shown in Figure 2.7 [10]. This is an extreme case where the adhesion between the encapsulant and the cell surface has failed across larger areas of the module.

Most of the modules in this system showed similar levels of delamination and subsequent corrosion of the metallization. Needless to say, these modules are no longer producing any power. The primary cause of delamination and then corrosion appears to be the use of EVA without primer. Only the glass was primed before lamination so the EVA did not adhere well to the cells.

Figure 2.7 Corrosion of the front metallization on solar cells [11]. *Source:* reprinted from author's PVSC paper.

Figure 2.8 Corrosion on corner of glass–glass module [11]. *Source:* reprinted from author's PVSC paper.

A second (less destructive) type of corrosion is shown in Figure 2.8 [11]. This is a glass–glass module where the non-EVA encapsulant did not adhere very well to the glass. Almost every night, water would condense in the delamination voids in the corner of this module. Ultimately, that liquid water corrodes the cells metallization locally. This array was actually deployed in Arizona so it shows that a humid environment is not required to corrosion to occur. In this particular case, less than 10% of the modules exhibited any corrosion at all.

The main way to prevent corrosion of the cell metallization is to prevent encapsulant delamination. Long-term adhesion of the encapsulant to all of the other components of the module is critical. The second most important design is the use of a moisture/corrosion

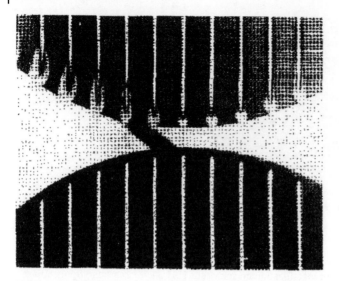

Figure 2.9 Corrosion of cells in Polyvinyl butyral (PVB) Module [12]. *Source:* reprinted from PVSC with permission of Ronald Ross of the Jet Propulsion Laboratory.

resistant metallization system. The screen printed pastes used today are much more tolerant of moisture than the materials that were available 20 years ago. When damp heat tests were first introduced, modules had trouble meeting the pass/fail criteria of a 5% power loss. Today, most commercial cry-Si modules exhibit no power loss after 1000 hours of damp heat testing.

The final design protection against metallization corrosion is the use of an encapsulant that retains low ionic conductivity when wet. In the early 1980s, module manufacturers tried to use Polyvinyl butyral (PVB) as an encapsulant as it was (and still is) used extensively in the laminated glass industry. The problem with PVB is that its ionic conductivity increases in higher humidity while that of EVA does not [12]. Cry-Si modules made with screen-printed metallization and PVB encapsulant failed in the field as the Ag from the screen-printed metallization migrated into the encapsulation as shown in Figure 2.9.

2.5 Encapsulant Discoloration

Discoloration of EVA encapsulant has been observed as a major cause of PV module power degradation. A typical example of the observed discoloration is shown in Figure 2.10. This module was fabricated using the original STR EVA formulation A9918 (standard cure EVA) and deployed on an open rack mount system at the STAR facility in Tempe, AZ as part of the control in an experiment to evaluate the susceptibility of different EVA formulations to discolor [13].

The initial observation and worst reported case of EVA discoloration occurred in the early 1990s at the Arco Solar Carrizo Plains site in California [14]. The Carrizo Plains system employed mirror enhancement so the modules were exposed to more UV light and

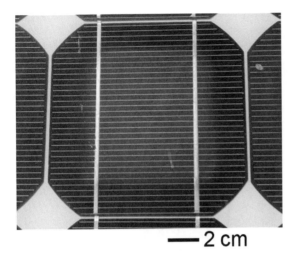

— 2 cm

Figure 2.10 Example of discolored Ethylene Vinyl Acetate (EVA) formulation A9918 in the STR experiment at the Tempe site [13]. *Source:* reprinted from author's PVSC article with the permission of David Miller of NREL.

higher temperatures than in standard open rack mounted terrestrial systems. The Carrizo system suffered severe power loss, originally thought to be entirely due to EVA browning. However, the author's team at Solarex was able to restore much of the lost power to several modules by repairing back solder bonds [15]. Much of the power loss in the system was due to solder bond issues, but the discoloration in these modules did lead to power loss in the order of 10–20%.

Additional observations confirmed both that the A9918 formulation of EVA did discolor and that the discoloration was caused by UV exposure at higher temperatures. At NREL, we had the opportunity to evaluate modules with A9918 EVA after 27 years of exposure at SMUD (Sacramento Municipal Utility District). Current losses of 10–12% were measured. This should be typical of the level of degradation expected for long-term exposure of this type of module. So it is not enough to invoke the warranty (usually based on 20% degradation) but it is a significant reduction allowing for much smaller reductions from all degradation modes before reaching the warranty limit.

Figure 2.11 provided by Govindasamy Tamizhmani of ASU shows images of two modules from the same module manufacturer where the module in images (Figure 2.11a) has been stored indoors (not exposed to the terrestrial environment) and the module in images (Figure 2.11b) has been exposed in the field for 10 years in Florida. Let's compare the two sets of pictures:

- Visual Image: In the picture of the unexposed module, no visual defects are observed. In the picture of the exposed module, there is no major discoloration, but with closer inspection some discoloration is observable in the centers of the cells.
- EL (Electroluminescence Image): In the EL picture of the unexposed module, a few darker spots are visible across the fingers. In the EL picture of the exposed module, the EL intensity is lower than for the unexposed module. Bright spots at cell interconnects probably correspond to solder bond degradation which will be discussed in Section 2.6.

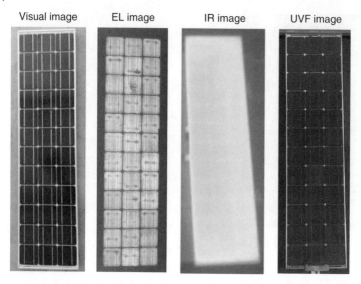

Figure 2.11a Characterization of unexposed module. *Source:* images provided by Govindasamy Tamizhmani of ASU.

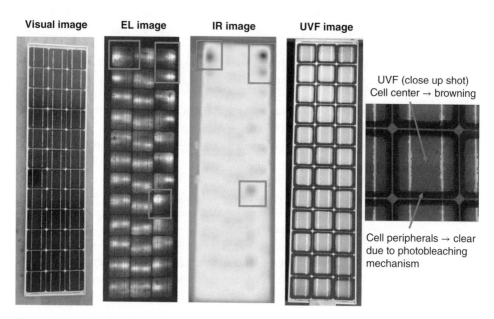

Figure 2.11b Characterization of module after 10 years of field exposure in Florida. *Source:* images provided by Govindasamy Tamizhmani of ASU.

- <u>IR (Infrared) Images (See Section 5.5 for a discussion of using IR as a tool to evaluate changes in PV modules.):</u> The IR image of the unexposed module has a uniform response with no hot spots. The IR image of the exposed module has hot spots in the same places that the EL image had bright spots, namely where solder bond degradation has occurred.

- UVF (UV Florescence) Images (See Ref. [16] for a description of how to use UVF as a tool to evaluate degradation in PV encapsulants.): The UVF image of the unexposed module shows no signal from the encapsulant. The UVF image of the exposed module shows a bright signal from the center of the cells with almost no signal from the edges of the cell. The bright areas indicate degradation of EVA material in the center of the cells, exactly where we see discoloration.

STR was able to replicate the discoloration and subsequent loss in short-circuit current via exposure to UV in a Weather-Ometer [17]. The Springborn Testing & Research Inc. (STR) work demonstrated that use of 15295 EVA formulation and cerium oxide in the glass achieves a combined reduction in the discoloration rate by a factor of 12.5 from the A9918 with standard low-iron glass [17, 18]. STR's experiments with the different additives within the A9918 package determined that the discoloration was not related to the EVA resin itself but was related to the additives, most importantly, the antioxidant (Naugard P) and its interaction with the cross-linking agent (Lupersol 101).

In the 1990s, the immediate solution for the PV industry was to use fast-cure formulations with cerium oxide glass for a superstrate. Over time, EVA manufacturers developed better formulations that virtually eliminated the discoloration problem. However, as we will see later in Chapter 6, loss of short-circuit current is still the dominate mechanism for power degradation in fielded cry-Si modules [19].

2.6   Failure of Electrical Bonds Particularly Solder Bonds

One of the major failure modes observed in the field is failure of the electrical bonds particularly solder bonds within the module. Electrical bonds can fail due to the stresses induced by thermal cycling or mechanical vibrations. PV modules are particularly susceptible to this because of the high currents flowing through the bonds during normal operation. So the modules themselves are heated by the sun while the bonds have an additional heat load from the current flow.

Early modules typically had only one solder bond at each end of the interconnect ribbon and only one ribbon per cell so failure of one solder bond resulted in an open circuit failure of the whole module. The tab across design used in most cry-Si modules results in multiple solder bonds on both the front and back of each cell providing redundancy for long-term survival in the field. Even today, non-cell solder bonds often have little or no redundancy so failure of one or two of these bonds can lead to the drop out of a cell string, a whole module or even a whole string of modules. Such single-point failures like that shown in Figure 2.12 are often workmanship related.

Most bus bar to cell solder bonds are made using automated equipment. In this case, process control and maintenance of the equipment is critical to achieving quality bonds. Since there is significant redundancy, failure of a few percent of those bonds will not affect long term performance. However, there are occasionally examples of large-scale failures within a product as shown in Figure 2.13. In this case, new production equipment was utilized, but it turned out the soldering equipment was not producing large enough solder bonds. This example will be discussed further in Chapter 4 as this particular failure mode required a change in the qualification testing method.

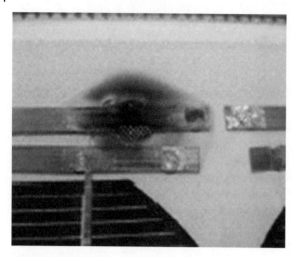

Figure 2.12 Example of single point solder bond failure on bus bar [1]. *Source:* reprinted from author's PVSC article.

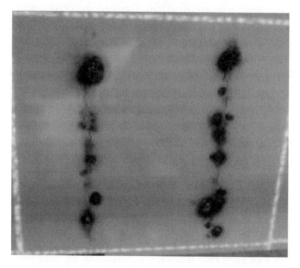

Figure 2.13 Example of Multiple Solder Bond Failures on One Cell [20]. *Source:* reprinted from author's EU PVSEC paper.

Over the years, the PV industry has made great strides in its efforts to alleviate solder bond failures including:

- Utilize multiple solder bonds on each tabbing ribbon.
- Utilize softer ribbon so there is less stress on the bonds themselves.
- Perform preventive maintenance on the soldering equipment to ensure a controlled process.
- Perform periodic pull tests to ensure quality of solder bonds being made.
- Perform thermal cycle tests well beyond the 200 cycles from the qualification tests.

- Implement training and QA inspections to ensure that non-cell solder bonds are being fabricated correctly.
- In critical areas (like termination wires) use both solder and mechanical connections.

2.7   Glass Breakage

We are all familiar with the fact that glass breaks if stressed enough. Heat strengthening or tempering does make glass stronger and less susceptible to breakage. The glass in cry-Si modules is usually tempered or heat strengthened. Such glass is usually strong enough to withstand the bending and flexing encountered during normal operation. However, there are still instances where this stronger glass breaks, including:

- High impact like a rock or a bullet will break it. Can almost always identify spot where the object hit. (See Figure 2.14.)
- Failure of or misuse of the support structure can lead to glass breakage. Pictures of this type of breakage will be shown in Section 2.10 on Structural Failures.
- High temperature (hot spot or arc) can also break glass. (See Figure 2.15 where you can readily see that the discolored area at the bus bar solder joint is the source of the break.)
- Improper handling, shipping and installation can also result in broken glass.

It is clear from both Figures 2.14 and 2.15, that when tempered glass breaks, the whole piece shatters into small fragments at the same time. Most of the time, a cry-Si module continues to work for a time after the glass breaks. However, over time the module suffers stresses like thermal cycling, wind loading and penetration of water to cause corrosion. Once the glass is broken the module lifetime is limited.

Annealed glass (similar in strength to the windows in your house) is not as strong as heat strengthened or tempered glass and the break grows across the surface in a random pattern. Thin-film modules often use annealed glass because the temperatures required for

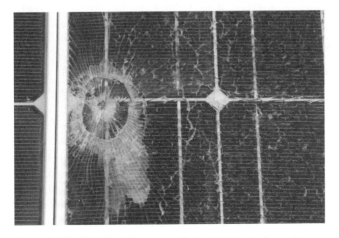

Figure 2.14   Module with tempered glass broken by impact.

Figure 2.15 Module with tempered glass broken due to overheating at bad solder bond.

depositing the thin-film semiconductors on the glass are high enough to anneal the heat strengthening out of the glass. Annealed glass breaks from the same stresses listed above for heat strengthened or tempered glass, but can also break due to:

- Thermal gradients: This is particularly important for thin-film PV because they are designed to absorb most of the light (not let it through like most glass). So shading such as provided by partial snow cover can result in thermal gradients that are large enough to break the glass. Solarex provided several kilowatts of a-Si Millenia modules to NREL for deployment in an array on the roof of their Outdoor Test Facility. The array was doing fine until there was a snow storm followed by a bright sunny day. The following week, it was observed that every single one of the Millenia modules had cracked glass. At Solarex, we then performed some experiments and validated that the annealed glass of the Millenia modules would crack with a temperature gradient in excess of 25° C from center to edge.

- Stress induced during processing (lamination): Lamination can result in edge pinch where the lamination process forces some of the encapsulant out of the corners and deforms the corners of the glass inward. Because this glass is under stress it tends to break soon after deployment.

- Stress from the mounting system: Since the annealed glass is not as strong as the tempered glass, it can break if installed under stress.

- Stress from handling: Any defects introduced into the glass during handling such as edge chips will ultimately result in breakage starting at the site of the defects.

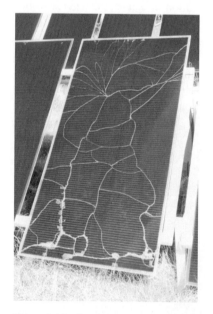

Figure 2.16 Breakage of an annealed glass thin-film module [20]. *Source:* reprinted from author's EU PVSEC paper with permission of David Miller.

Figure 2.16 shows a picture of a broken thin-film module made with annealed glass.

Glass breakage can be a major failure mode for thin-film modules especially during installation. On the other hand, it is usually not a major issue for cry-Si because of the use of tempered glass. However, even with tempered glass there have been several reported cases where most of the modules in an array were broken. In one case, the culprit was a high wind that picked up the rocks used for ballast on the roof and smashed them into the modules. The modules didn't break immediately but over the next few weeks almost all of them suffered broken front glass. Other examples have usually involved hail storms with very large hail stones.

2.8 Junction Box Problems

Junction boxes would appear to represent a mature technology yet most reports of PV field failures include junction box failures as one of the most prevalent. Junction boxes can fail in the field by:

- Becoming detached from the backsheet as shown in Figure 2.17. In the figure, the box has moved from its original position most likely compromising the seal and allowing liquid water into the box leading to corrosion of the electrical connections. In other cases, boxes have been observed to fall off the module completely. Poor adherence of the box can be attributed to either selection of the wrong adhesive or poor workmanship. Some adhesives, particularly contact adhesives like tape, can withstand a short-term pull at room temperature (a test included in module qualification – See Chapter 4) but peel apart under long-term stress at elevated temperatures. Obtaining a waterproof seal around the box requires careful deposition of the adhesive around the perimeter of the box which requires good process control.
- Condensation of water within the box. Many encapsulants absorb moisture from the air. When the module cools quickly at night the moisture is driven out of the encapsulant. The easiest place for it to go is to condense within any open spaces particularly within the junction box. So if the box is hermetically sealed and not potted it may fill with water

Figure 2.17 Junction box that has moved from its mounting position. *Source:* picture provided by David Miller of NREL.

from the encapsulant leading to corrosion. Solutions to this problem include leaving a weep hole in the box or potting the box.

- Failure of the junction box lids. To save on the labor cost of attaching junction box lids many are designed as snap fits. A number of these designs have failed in the field either popping open or cracking. The stresses created by the large temperature differences observed in the field are the likely cause of these failures. Once again, any such design should be adequately validated using accelerated stress tests (See Chapters 3 and 4) before being deployed.

Corrosion of the electric connections within the junction box is both a performance and safety issue. Failure of the output connections from a module can easily result in loss of power for a whole string of modules. In the process of corrosion, the electrical connections have the potential to cause a dc arc, which will be discussed in section 2.14.

2.9 Loss of Elastomeric Properties of Back Sheets

Long-term exposure of polymers to sunlight and high temperature can cause some materials to degrade.

Because these polymers usually provide the electrical insulation between the active cell circuit and the outside world (e.g. backsheet), their failure can result in potential safety hazards. Figure 2.18 shows an example of a backsheet that has cracked due to UV exposure. Typically, in an open rack configuration, the rear side of the backsheet can expect to see up to 20% of the UV that falls on the front while some UV passing through the front. At the elevated temperatures at which the modules operate, this can be sufficient to degrade materials that are not UV stable.

During the 1980s and 1990s most backsheets were made using outer layers of Tedlar, a Polyvinyl Fluoride film that is quite UV stable. However, because of the higher cost of Tedlar, module manufacturers began using alternate materials for backsheets. Some of

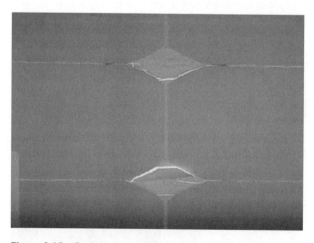

Figure 2.18 Backsheet exhibiting cracking after outdoor exposure. [3]. *Source:* reprinted from author's PVSC article.

these alternative materials have cracked during field exposure [21]. Such cracks are potential safety hazards, especially if the modules are used in high-voltage systems. On the other hand, a number of fielded modules have been observed to lose backsheet integrity without serious loss in performance [11]. While it is clearly better to use a backsheet material that will survive for the life of the module, degradation of the backsheet may not result in loss of performance of the modules. This may be okay if these modules are used in lower voltage systems or behind the fence where personnel can deal with any potential danger.

2.10   Reverse Bias Hot Spots

Reverse Bias Hot-spot heating occurs in a module when its operating current exceeds the reduced short-circuit current (Isc) of a shadowed or faulty cell or group of cells. When such a condition occurs, the affected cell or group of cells is forced into reverse bias and must dissipate power. If the power dissipation is high enough or localized enough, the reverse biased cell can overheat resulting in melting of solder and/or silicon, deterioration of the encapsulant and backsheet and even breakage of the glass as shown in Figure 2.19. A hot spot can be a localized phenomenon that results in a small region of the cell melting and thus shorting the whole cell thereby removing the hot spot but also rendering the cell incapable of producing power. In more severe cases, enough heat is dissipated to melt the encapsulant and backsheet resulting in a breach of the module's electrical insulation; a safety hazard. In the most severe cases, the module can catch fire and the glass can break. Therefore, modules should be designed and constructed to minimize the potential for reverse bias hot-spot formation.

The susceptibility of solar cells to reverse bias hot spots depends upon their reverse bias characteristics. The reverse bias characteristics of solar cells can vary considerably [23]. Cells can either have high-shunt resistance where the reverse performance is

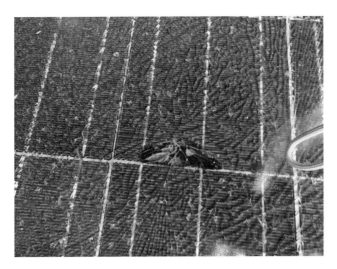

Figure 2.19   Glass breakage due to hot spot in module [22]. *Source:* reprinted from author's PVSC article.

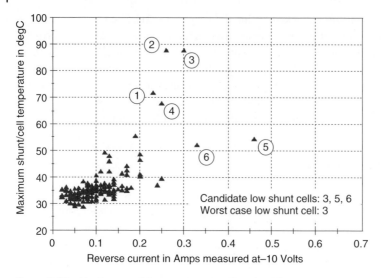

Figure 2.20 Maximum cell temperature as a function of reverse current measured at 10V reverse bias. [24]. *Source:* reprinted from author's PVSC article with the permission of Werner Herrmann of TÜV Rheinland.

voltage-limited or have low-shunt resistance where the reverse performance is current-limited. Each of these types of cells can suffer hot spot problems, but in different ways.

Low-Shunt Resistance Cells:

- The worst case shadowing conditions occur when the whole cell (or a large fraction of it) is shadowed.
- Often low-shunt resistant cells are this way because of localized shunts. In this case hot spot heating occurs because a large amount of current flows in a small area. Because this is a localized phenomenon, there is a great deal of scatter in performance of this type of cell as shown in Figure 2.20. Cells with the lowest-shunt resistance have a high likelihood of operating at excessively high temperatures at least in localized areas when reverse biased as shown in Figure 2.21.
- Because the heating is localized, hot spot heating of low-shunt resistance cells occurs quickly.

High-Shunt Resistance Cells:

- The worst case shadowing conditions occur when a small fraction of the cell is shadowed.
- High-shunt resistant cells limit the reverse current flow of the circuit and therefore heat up. The cell with the highest shunt resistance will have the highest power dissipation.
- Because the heating is uniform over the whole area of the cell, it can take a long time for the cell to heat to the point of causing damage.

Either type of cell can suffer from reverse bias hot spot heating. To protect the cells from reverse bias hot spots, modules are designed and built with incorporation of bypass diodes. A bypass diode will turn on and conduct current when its voltage is in the forward direction.

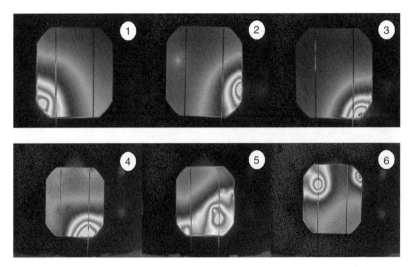

Figure 2.21 IR pictures of heating effects and the formation of hot spots due to increased reverse voltage on six different cells [24]. *Source:* reprinted from author's PVSC article with the permission of Werner Herrmann of TÜV Rheinland.

This occurs when the voltage drop across the shadowed cell (or cells) exceeds the forward voltage of the remaining cells in series with it that are protected by that diode. The number of cells protected by an individual diode determines the level of reverse voltage that can appear across a shadowed cell. In cry-Si modules it is fairly typical to incorporate one bypass diode for every 20–24 cells. This means that a shadowed cell should see no more than about 10–12 V reverse bias under worst-case conditions. If all of the cells can withstand 10 or 12 V reverse bias without having hot spot issues this design is fine. There are, however, cases where cells cannot tolerate 10 or 12 V reverse bias. In some cases, this is due to localized shunts. In at least one case, the old BP Solar Saturn product with buried contact cells, the cell shunt resistance was so high that shadowed cells reached dangerously high temperatures unless the number of cells per diode was reduced to 12.

The safety of PV modules to reverse bias hot spot problems is validated via use of the Hot Spot Test that has been incorporated into the Module Qualification Test and will be discussed in Chapter 4.

2.11 By-Pass Diodes

Section 2.10 explained that bypass diodes are an important addition to PV modules to alleviate potential damage from reverse bias hot spots. These diodes also protect the module from internal arcing from an open circuit that will be discussed in Section 2.14. Of course adding any additional component to the module has the potential to add new failure modes. Failure of bypass diodes has been reported in a number of PV systems. Kato [25] reported that 593 diodes had failed in 1272 modules that were inspected from roof-mounted systems in Japan. Tamizhmani [26] reported on 26 failed diodes from 2352 modules inspected in a utility-scale power plant in Arizona. Figure 2.22 shows an example of

Figure 2.22 Picture of failed diodes. *Source:* picture provided by Govindasamy Tamizhmani of ASU.

a failed diode from Tamizhmani. Modules with bad bypass diodes are no longer protected against hot spots so can be damaged by something as simple as a bird leaving something behind on the module.

Bypass diodes can fail in either short-circuit or open-circuit mode. If it fails short circuited, it will pass current whenever the sun is shining so the cells it is protecting will no longer produce any useful electricity. If the diode fails open circuit it acts as though it were not in the electric circuit at all and therefore provides no reverse bias hot spot protection to the module. It is fairly easy to see when a diode short circuits as the module no longer produces the rated amount of electricity and the diode is on and hot whenever the sun shines. On the other hand, an open-circuited diode is difficult to observe since it has no impact on an operating module. To determine whether bypass diodes are operating or not requires a module-by-module shading test that is very time consuming.

Bypass diodes fail due to a number of different causes:

1) From static electricity during insertion of the diodes into the module – in this case, the module gets shipped to the customers with bad diodes.
2) By overheating due to either poor diode selection or inadequate cooling of the diode.
3) Due to thermal runaway when the diode switches from the conducting to the non-conducting state, once again due to either poor diode selection or inadequate cooling.
4) From failure of the diode connections due to thermal and/or mechanical cycling.
5) From connecting the module into the circuit backwards especially hooking it to a battery in the wrong polarity.
6) From lightening striking near the module as the bypass diodes are more susceptible to lightning strikes in the vicinity of the modules than the solar cells.

The qualification test, discussed in Chapter 4, contains a Bypass diode thermal test that directly addresses overheating, item 2 above. But because diodes can fail in so many different ways there is no simple test that can tell you whether a particular diode protection scheme will survive for the life of the module. The following is a list of potential solutions for the different issues listed above. This work is now ongoing and will be discussed later in the book in the appropriate chapters.

1) To develop a standard that measures the electrostatic voltage encountered while inserting the diodes into the module so manufacturers can take the necessary precautions. In addition, the development of a quality management system for PV modules that contains recommendations on how to determine that the diode is in place and functional at the time of shipment. (See Chapter 7.)
2) The qualification test includes a Bypass Diode Thermal Test that went a long way toward eliminating use of inadequate or poorly heat sunk diodes. The PVQAT diode team is assessing whether a revision of the bypass diode thermal may be necessary [27] to take into account higher module operating temperatures especially on roof tops. (See Chapter 7.)
3) The PVQAT diode group is also developing a test to evaluate a diode's potential for thermal runaway during switching [28]. (See Chapter 7.)
4) To address diode connection failures due to thermal and mechanical cycling, the latest edition of the qualification test requires that the bypass diodes be functional after the end of the accelerated stress tests. (See Chapter 4.)
5) Proper installation of modules is always important. This is being addressed by the IECRE System and a quality management system for installation of PV systems. (See Chapter 8.)
6) To address lightning strikes, we take the usual system level precautions and then pray that lightening doesn't strike our PV system.

2.12 Structural Failures

Sometimes the module itself is not strong enough to survive mechanical stresses, particularly those from wind and snow load. Such failures can involve the superstrate, substrate, frame and/or frame adhesive. Often, it is the way the module is mounted that determines whether it can survive a particular load. Figures 2.23a and Figure 2.23b shows the same module type undergoing snow load testing from IEC 61215 (See Chapter 4.) while mounted

Figure 2.23a Two modules of same type undergoing snow load testing [29]. Module supported on the ends.

Figure 2.23b Two modules of same type undergoing snow load testing [29]. Module supported at the design points. *Source:* reprinted from author's EU PVSEC paper.

Figure 2.24 Snow damaged modules hanging from their support structure by electrical cables [20]. *Source:* reprinted from author's EU PVSEC paper.

in two different ways [29]. In Figure 2.23a, the module is mounted on the ends. Notice how much the module has deflected under the snow load test. In Figure 2.23b, the module is mounted at the design points. As you can see, there is virtually no deflection of the module in this case. Which do you think is more likely to survive many harsh winters in a snowy location?

The modules in Figure 2.24 have been damaged by snow load [20]. A spring snowstorm dumped several feet of heavy wet snow on top of the solar canopy. The mounting system was not strong enough so the frames twisted and the laminates slid out. When I took the picture, the laminates were hanging over the parking lot, held only by their electrical cables.

The module in Figure 2.25 has also been damaged by snow load. The force of the snow sliding down the module has pushed the bottom frame off the glass superstrate. Once the glass was not supported by this front frame member, it flexed too much and broke.

Figure 2.25   Snow load damage to module.

The static mechanical load test included in qualification is not designed for, nor capable of, determining if the type of failure seen in Figure 2.25 can occur. For this a new "non-uniform snow load test" is under development in IEC.

Often it is the way the module is mounted that determines whether it can survive a particular load. There are a number of examples that you do not want to repeat including solar arrays held down using ballast that have blown over and modules mounted on roof tops that have blown off. Some recommendations for avoiding structural failures:

- Follow the module manufacturer's installation instructions. This may seem trivial but there are many examples of installers using mounting methods that did not follow the mounting instructions of the module manufacture and ultimately failed. It should be noted at this point, that if you do not mount a safety certified module as recommended by the manufacturer, the safety certification is no longer valid.
- Follow published standards for modules and mounting systems.
- Follow local building codes where available as they are usually based on local weather history. Local building codes are based on what can be expected at a particular location so the code should be based on the expected highest wind and snow loads.
- Test new approaches (for example, in a wind tunnel) before using them in the field.

Finally, a system that is installed improperly may be unsafe and unreliable even if it is well designed and uses the highest quality components. Installer training and certification programs are critical to achieving highly reliable and safe PV systems. (See Chapter 8.)

2.13   Ground Faults and Open Circuits Leading to Arcing

A high voltage difference between two points separated only by a small air gap can lead to an electric arc. While arcs can occur in ac and dc circuits, dc arcs are less well understood and tend to survive longer (no zero cross-over of voltage). In the wrong situation, dc arcs can be sustained for a significant amount of time and because they are so hot they can cause almost any material in the vicinity to catch fire. Because of the requirement for high-voltage, individual modules and even low-voltage systems are not likely to suffer from arcing. On the other hand, the newer very high voltage systems (> 1000 V) are much more likely to suffer from arcing problems.

Three types of arcs can occur in PV modules:

- Series – A series arc occurs when a connection open circuits while the module is producing current. Any intermittent connection in the module has the potential for producing a dc arc fault. These connections may include soldered bonds within the module, the leads bringing the power out of the laminate or the actual connectors that are commonly used on the wire leads attached to PV modules.
- Parallel – Parallel arcs occur when two conductors of opposite polarity in the same dc circuit come into contact. The voltage within most modules is usually too low for a parallel arc to occur. In PV, parallel arcs are more likely to occur in the wiring where higher voltages are often routed through dual cable wires. One potential cause of such wire failure is rodents chewing through cables.
- To ground – This fault requires the failure of the module insulation system in either one place if the array is already grounded or two places if the array is ungrounded. (In an ungrounded array the first ground fault grounds the array.) Typically, the ground fault paths are not capable of carrying the high currents found in PV systems so the pathways overheat, leading to an arc. While a GFDI (Ground Fault Detector & Interrupter) provides some measure of protection against this fault, there have been cases of faults to ground that failed to trip the GFDI protection yet created an arc.

Since we are mainly concerned with modules in this book, we will consider only series arcs caused by open circuits and ground faults.

Figure 2.26 shows a picture of an active arc caused by an open circuit within the junction box of a module. This arc was caused on purpose in the test field at NREL. A string of 5–200 watt cry-Si modules was short circuited. Then one of the cable glands was loosened and the cable was pulled. The arc started quickly and burned until the wire feedthrough

Figure 2.26 Arc caused by an open circuit within a junction box.

Figure 2.27 Module with ground fault to frame arc [11]. *Source:* reprinted from author's PVSC paper.

melted and the input cable fell to the ground. While the arc was active, everything in contact with it burned, including the junction box that was rated as fire resistant. Once the arc stopped the fire went out quickly. (The electronic version of this book contains a movie of this whole experiment.)

The module terminations are probably the most likely places to see a series arc within a PV module since a single-point open circuit can cause the arc. Within the module laminate, it usually takes multiple failures to cause an arc. Even if all of the interconnect ribbons connecting two adjacent cells open up (either because the ribbons themselves break or all the solder bonds to one of the cells fail, a functional bypass diode can still carry the current around the open circuit, preventing an arc). On the other hand, many thin-film modules do not use bypass diodes, so broken superstrate glass can lead to arcing in a monolithic thin-film module.

Figure 2.27 shows a picture of a module that suffered a ground fault to the frame [11]. This ground fault must have continued for a number of days, starting each morning and stopping each evening, as it had burned a path of more than 30 cm along the edge of the module. This module used a soft silicone encapsulant and was built before automation, so the module was likely manufactured with the cells too close to the frame. This module was designed and built before the module qualification test IEC 61215 was published so the design probably did not undergo the extensive dielectric withstand and wet high pot tests that are utilized today.

In some cases, multiple ground faults are the result of poor installation practices. In Figure 2.28, the installer mounted the modules with clips that penetrated the module insulation and contacted the solar cells at numerous locations. As soon as the system was hooked up and turned on it caught fire. Luckily, it was extinguished before the fire spread to the house. Mounting modules with conductive hardware on polymeric backsheets or frontsheets within the active cell area should be avoided or done with extreme caution.

Figure 2.28 Array fire caused by ground faults in the installation system. *Source:* picture provided by Daniel Cunningham.

2.14 Potential Induced Degradation

Potential Induced Degradation (PID) refers to a type of module degradation that is due at least in part to the voltage of the PV system. So, typically the degradation only occurs on modules in one polarity and the degradation rate increases with increasing voltage. Over the years, several different types of PID have been reported. The first example occurred back in the JPL Block buy days when Arco Solar tried to use PVB as their encapsulant. As JPL explained [30, 31], the conductivity of PVB increased by several orders of magnitude when it was saturated with water vapor. In high humidity areas at high voltages, silver ions were pulled out of the grid lines and migrated through the PVB toward the grounded frame resulting in significant degradation of the series resistance and therefore the power of the modules. Figure 2.9 showed migration of silver from the grid lines.

A second type of PID occurred in some of the earliest amorphous silicon modules. The initial Transparent Conductive Oxide (TCO) glass on which the a-Si material was deposited did not contain a barrier layer to stop sodium ion migration. When these early a-Si modules were used in high-voltage systems the voltage caused sodium ions to flow through the glass into the TCO leading to its delamination from the glass. An example is shown in Figure 2.29 [32]. Once the phenomenon was understood [33], the glass manufacturers were able to develop a barrier layer that kept the sodium from attacking the TCO. Subsequent a-Si modules were much more tolerant of PID but there is still evidence of some PID in the a-Si modules deployed in a number of larger systems including Springerville, AZ.

The next episode of PID occurred with a new type of cell, the SunPower back contact cell. These cells used a charged front surface to reduce carrier recombination and increase efficiency. When these modules were used in high-voltage systems, the modules at the positive end lost power over time due to leakage current through the glass and encapsulant resulting in a change in the charge at the front surface. SunPower published installation instructions

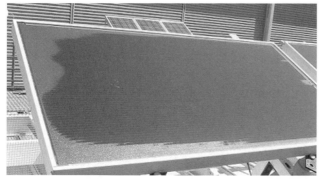

Exposed for 12 months at -600 v in FL

Figure 2.29 Electrochemical corrosion of TCO in a-Si module [32]. *Source:* reprinted from PVSC paper by Neelkanth Dhere from project funded by author while at Solarex.

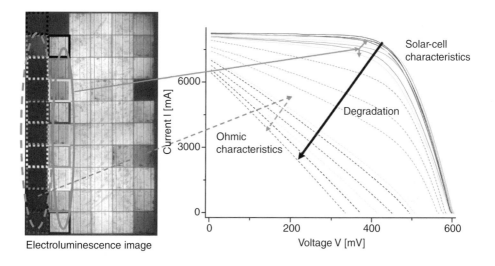

Electroluminescence image

PID-s → Potential induced degradation by shunting of solar cell

Figure 2.30 Electroluminescence (EL) picture of module after Potential-Induced Degradation (PID) degradation associated power loss [34]. *Source:* reprinted from NREL Module Reliability Workshop with permission of Max Koentopp of Hanwha Q Cells.

to use their modules in only positively grounded systems. This eliminated the problem until they eventually developed an improved cell process that did not suffer from this type of PID.

The PID that is of most concern today involves standard screen-printed cry-Si solar cells with silicon nitride (SiN) anti-reflective (AR) coatings. Almost the entire cry-Si PV industry has switched to SiN AR coatings as they result in higher cell efficiencies. The degradation involves flow of sodium ions through the front glass in modules mounted at high negative voltage. The sodium ions then shunt the solar cells. This can't be seen by the naked eye, but is easily observed using EL as shown in Figure 2.30 where the dark cells are shunted.

The degradation rate is worse when the surface of the glass is wet, either from dew or rain. PID-sensitive modules on the negative, high voltage end of the circuit suffer the most power loss. In some cases, the modules have lost most of their output power in less than one year exposure time [35]. In some cases, much of the power has been restored by reversing the polarity of the voltage.

This type of PID is a fairly new phenomenon because:

- Until recently, most PV systems were fairly low voltage and the majority of inverters used transformers with the negative side of the PV array grounded.
- In the US, the National Electric Code (NEC) required the grounding of one pole of the PV DC circuit. The grounded side was usually the negative one.

Now with the advent of transformerless inverters and changes to the NEC, there are many more systems being installed with modules at high negative voltage.

There are several ways to reduce or eliminate the effect of PID on PV modules made with SiN AR, screen printed cry-Si cells.

- Tune the index of refraction of the SiN AR coating.
- Use an encapsulant with lower conductivity.
- Use a non-conductive mounting system.

Initially, many cell manufacturers tuned the index of the AR on their cells to get a quick fix. While this can work well, the success depends on how well they can control the SiN deposition process. This may be why some module types labeled PID-resistant still occasionally suffer from PID as the factory process drifts. Long term, the use of a lower conductivity encapsulant is probably the most effective prevention method.

One of the reasons that PID has been such a big problem is the fact that the module qualification test sequence does not test for it. Hopefully, this will be remedied in the near future. Today, there is a published IEC test specification, IEC TS 62804-1 [36] that contains two methods for determining a module's sensitivity to PID.

2.15 Thin-Film Specific Defects

Several type of defects or failure modes are specific to thin-film PV modules, particularly monolithically integrated modules where the semiconductor is deposited onto a substrate, usually glass, and then patterned directly to form the series–parallel circuitry. These failure modes are related to the way the thin-film modules are manufactured or the way they are packaged in the module. Other types of thin-film module issues are similar to those encountered with cry-Si, like glass breakage and PID so have already been discussed. Table 2.3 contains a list of thin-film specific defects and failure modes. These will each be discussed in the following subsections.

2.15.1 Light-Induced Degradation

Light-induced degradation (LID) as seen in amorphous silicon is not really a failure or defect since it occurs in all the products at a well understood rate. The Staebler-Wronski-effect [37] results in a loss of cell/module efficiency on exposure to light.

Table 2.3 Failure modes specific to thin modules.

Light Induced Degradation (LID)

Inadequate Edge Deletion

Shunts at laser scribes

Shunts at impurities in films

Failure of Edge Seals

The amount of efficiency loss depends on the quality and thickness of the a-Si layers, the temperature, and the intensity of the incident light. These result in an initial power loss, typically in the order of 15%, followed by a seasonal variation in output performance. See, for example, reference [38] for a good explanation of the seasonal effects. The initial power loss means that the system installer must design the system and its BOS to work and be safe with both the initial and degraded power levels. Both the initial power loss and the seasonal variation must be taken into account when predicting system performance.

2.15.2 Inadequate Edge Deletion

For monolithically integrated thin-film modules, the thin-film semiconductor is deposited onto a piece of glass. The thin-film layers must then be removed from the edges of the glass plate. If the thin-film layers are not adequately removed from the edges, the remaining material can cause high-leakage currents and provide pathways for moisture ingress into the module. High-leakage currents can be a safety issue (causing personnel exposure to the PV system's voltage) and a reliability issue (electro-corrosion of TCO, contacts and thin-film PV layers as was discussed in Section 2.14).

The best way to determine whether the edge deletion has been successful and the leakage current is under control is to perform a wet leakage current test. The wet leakage current test was first introduced into the Solar Energy Research Institute Interim Qualification Test (SERI IQT) and eventually into the IEC 61646 Qualification Standard for thin films as will be described in Chapter 4. Of course, using this test on qualification only ensures that those modules delivered to the test laboratory actually had adequate edge deletion. In addition to improved edge deletion process control, some thin-film PV module manufacturers subject 100% of their modules to a production line wet leak current test to ensure adequate edge deletion.

2.15.3 Shunts at Laser Scribes and Impurities in Thin Film

Lasers are often used to delineate thin-film cells both for monolithic designs and for those using standard cells cut from larger substrates. Improper application of the scribe lines can lead to shunting of some of the cells. In addition, because the thin-film layers are so thin, impurities that are trapped in the films during manufacturing can lead to shunting. Regardless of what causes the shunting, the shunted region carries higher current than designed and results in higher temperature in the vicinity of the shunt. Higher temperatures can lead to localized discoloration of the encapsulant. Both the higher temperature

Figure 2.31 Hot spot in monolithic, thin-film module after field exposure [11]. *Source:* reprinted from author's PVSC article with the permission of David Miller of NREL.

Figure 2.32 Hot spot in discrete cell, thin-film module after field exposure. *Source:* picture provided by David Miller of NREL.

and the increased current flow can damage the thin-film cell resulting in localized power loss. Shunted regions have even been observed to get hot enough to break the glass.

Figure 2.31 shows an example of a localized discoloration caused by a shunt in a monolithic type module. Figure 2.32 shows an example of a localized discoloration caused by a shunt in a module with discrete thin-film cells. The major solution for controlling these types of hot spots is process control. That is to keep them from occurring in the first place. IR scanning can also be used to identify modules with potential shunting problems. In Chapter 7, we will come back to the shunting problem in thin-film modules in the discussion on the work of PVQAT.

2.15.4 Failure of Edge Seals

Many types of thin-film modules are sensitive to moisture ingress. Thin films of the different materials that make up the module, like TCOs, metal conductors or even the semiconductors themselves, may be directly sensitive to moisture or may suffer from electro-corrosion that is enhanced by increased conductivity when the packaging materials absorb moisture. Therefore, as described in Chapter 1, many thin-film modules use a hermetically sealed package consisting of glass superstrate, glass substrate and an edge seal. (Either of the two

Figure 2.33 Edge seal failure in field exposed thin-film module [11]. *Source:* reprinted from author's PVSC article with the permission of David Miller of NREL.

pieces of glass could be replaced by any solid moisture barrier in this construction.). For the module to survive, the edge seal must continue to keep moisture out of the package.

Do edge seals survive in the field and continue to perform their designed function for 25 or 30 years? The answer at this time is that we do not know. No PV modules with edge seals have been deployed for 25 years. There are certainly examples of edge seals that have failed in the field (See Figure 2.33.) [11] although these are from an earlier generation and one would expect that today's edge seals are more robust.

Today, most thin-film modules are qualified to IEC 61646 the thin-film qualification test sequence that has now transitioned into technology specific parts of IEC 61215 (See Chapter 4.). This test sequence contains a 1000 hour damp heat test, which according to Kempe [39] is not a long enough test to evaluate edge seal survival in many terrestrial climates. His work also points out that in addition to moisture diffusing through the edge seal, it can also penetrate down the glass/edge seal interface if the edge seal material does not maintain adequate adhesion to the glass. Based on this work, a new IEC standard for Edge Seal Durability is being developed. Once this methodology has been developed and perfected, we can expect the long-term performance of edge seals to be better understood.

References

1 Wohlgemuth, J., Kurtz, S., Sample, T. et al. (2014). Development of Comparative Tests of PV Modules by the International PV QA Task Force. 40th IEEE PVSC in Colorado, USA (8–13 June 2014).

2 Engineering ToolBox (2001). www.engineeringtoolbox.com (Accessed 28 August 2019).

3 Wohlgemuth, J. and Kurtz, S. (2015). Photovoltaic Module Qualification Plus Testing. 42nd IEEE PVSC in New Orleans, USA (14–20 June 2015).

4 IEC 62759-1 (2015). Photovoltaic (PV) modules – Transportation testing – Part 1: Transportation and shipping of module package units.

5 IEC 63049 (2017). Terrestrial photovoltaic (PV) systems – Guidelines for effective quality assurance in PV systems installation, operation and maintenance.

6 Rutschmann, I. (2012). Unlocking the Secret of Snail Trails. *Photon Magazine* (29 January), 114–125.

7 Duerr, J., Bierbaum, J.M., and Philipp, D. (2016). Snail Tracks: Identification of Critical Environmental Stresses, Corrosion Products and Influences of Module Components. *Proceedings of NREL PVMRW*, Colorado, USA (23 February 2016).

8 Wohlgemuth, J.H., Hacke, P., Bosco, N. et al. (2016). Assessing the Causes of Encapsulant Delamination in PV Modules. 43rd IEEE PVSC in Portland, USA (5–10 June 2016).

9 Bosco, N., Kurtz, S., Tracy, J., and Dauskardt, R. (2016). Development and First Results of the Width-Tapered Beam Method for Adhesion Testing of Photovoltaic Material Systems. 43rd IEEE PVSC in Portland, USA (5–10 June 2016).

10 Miller, D.C., Annigoni, E., Ballion, A. et al. (2016). Degradation in PV Encapsulant Adhesion: An Interlaboratory Study Towards a Climate-Specific Test. 43rd IEEE PVSC in Portland, USA (5–10 June 2016).

11 Wohlgemuth, J., Silverman, T., Miller, D.C. et al. (2015). Evaluation of PV Module Field Performance. 42nd IEEE PVSC in New Orleans, USA (14–20 June 2015).

12 Mon, J., Wen, L., and Ross, R. (1987). Encapsulant Free-Surfaces and Interfaces: Critical Parameters in Controlling Cell Corrosion. 19th IEEE PVSC in New Orleans, USA (1–4 May 1987).

13 Wohlgemuth, J.H., Kempe, M.D., and Miller, D.C. (2013). Discoloration of PV Encapsulants. 39th IEEE PVSC in Florida, USA (16–21 June 2013).

14 Rosenthal, A.L. and Lance, G.C. (1991). Field test results for the 6 MW Carrizo solar PV power plant. *Solar Cells* 1–4: 563–571.

15 Wohlgemuth, J. and Petersen, R. (1993). Reliability of EVA modules. 23rd IEEE PVSC in Louisville, USA (9–14 May 1993).

16 Eder, G., Voronko, Y., Kubicek, B., and Knoebl, K. (2017). Fluorescence Spectroscopy and Imaging on aged polymeric PV-Encapsulantes. European Symposium of Polymer Spectrocopy, Dresden. https://www.researchgate.net/publication/308522512_ Fluorescence_Spectroscopy_and_Imaging_on_aged_polymeric_PV-Encapsulantes.

17 Tucker, R.T., Kuitche, J.M., Arends, T. et al. (2006). Nine (9)-Year Review of Field Performance of EVA-Based Encapsulants. 21st EU PVSEC in Dresden, Germany (4–8 September 2006).

18 Holey, W.W. and Agro, S.C. (1998). *Advanced EVA-Based Encapsulants*. Colorado, USA: NREL.

19 Jordan, D.C. and Kurtz, S.R. (2013). Photovoltaic degradation rates – an analytical review. *Prog. PV* 21 (1): 12–19.

20 Wohlgemuth, J.H., Kurtz, S.R., Miller, D.C., and Bosco, N.S. (2015). *PV Module Reliability: How Can we Improve it?* EU PVSEC.

21 John Trout, T. (2018). *Understanding PV Module Durability through Analysis of Fielded Modules and Sequential Accelerated Testing*. NREL PVRW.

22 Wohlgemuth. J.H. and Kurtz. S.R. (2012) How can we make PV Safer?, 38th IEEE PVSC, in Austin, Texas (3-8 June, 2012).

23 Herrmann, W., Alonso, M., Boehmer, W., and Wambach, K. (2001). Effective Hot-Spot Protection of PV Modules – Characteristics of Crystalline Silicon Cells and Consequences for Cell Production. 17th EU PVSEC in Munich, Germany (22–26 October 2001).

24 Wohlgemuth, J. and Herrmann, W. (2005). Hot Spot Tests for Crystalline Silicon Modules. 31st IEEE PVSC in Florida (3–7 January 2005).

25 Kato, K. (2012). *PVRessQ, PV Module Failures Observed in the Field*. NREL PVMRW.

26 TamizhMani, G. (2014). *Reliability Evaluation of PV Power Plants*. NREL PVMRW.

27 Shiradkar, N., Gade, V., Schneller, E.J., and Sundaram, K. (2015). Revising the Bypass Diode Thermal Test in IEC 61215 Standard to Accommodate Effects of Climatic Conditions and Module Mounting Configurations. 42nd IEEE PVSC in New Orleans, USA (14–19 June 2015).

28 Shiradkar, N., Schneller, E., Dhere, N., and Gade, V. (2014). Predicting Thermal Runaway in Bypass Diodes in PV Modules. 40th IEEE PVSC in Colorado, USA (8–13 June 2014).

29 Wohlgemuth, J. H., Cunningham, D. W., Amin, D. et al. (2008). Using Accelerated Tests and Field Data to Predict Module Reliability and Lifetime. 23rd EU PVSEC in Valencia, Spain (1–5 September 2008).

30 Mon, G., Wen, L., and Ross, R. (1987). Encapsulant Free-Surfaces and Interfaces: Critical Parameters in Controlling Cell Corrosion. 19th IEEE PVSC in New Orleans, USA (1–4 May 1987).

31 Mon, G., Orehotsby, J., and Ross, R. (1984). Predicting Electrochemical Breakdown in Terrestrial Photovoltaic Modules. 17th IEEE PVSC in Florida, USA (1–4 May 1984).

32 Dhere, N.G., Hadagali, V.V., and Jansen, K. (2005). Performance Degradation Analysis of High-Voltage Biased Thin-Film PV Modules in Hot and Humid Conditions. 31st IEEE PVSC in Florida, USA (3–7 January 2005).

33 Carlson, D.E., Romero, R., Willing, F. et al. (2003). Corrosion effects in thin-film photovoltaic modules. *Prog. PV Res. Appl.* 11: 377–386.

34 Koentopp, M.B., Schulze, A., and Taubitz, C. (2016). Potential Induced Degradation (PID): A Physical Model Describing Degradation and Recovery in the Field, NREL PVMRW.

35 Pingel, S., Frank, O., Winkler, M. et al. (2010). Potential Induced Degradation of solar cells and panels. 35th IEEE PVSC in Hawaii, USA (20–25 June 2010).

36 IEC TS 62804-1 (2015). Photovoltaic (PV) modules – Test methods for the detection of potential-induced degradation – Part 1: Crystalline silicon.

37 Staebler, D.L. and Wronski, C.R. (1977). Reversible conductivity changes in discharge-produced amorphous Si. *Appl. Phys. Lett.* 31: 292.

38 Gottschalg, R., Jardine, C.N., Rüther, R. et al. (2002). Performance of Amorphous Silicon Double Junction Photovoltaic Systems in Different Climatic Zones. 29th IEEE PVSC in New Orleans, USA (17–24 May 2002).

39 Kempe, M.D., Dameron, A.A., and Reese, M.O. (2013). Evaluation of moisture ingress from the perimeter of photovoltaic modules, *Prog in PV: Res. Appl.* https://DOI:10.1002/pip.2374,2013.

3

Development of Accelerated Stress Tests

Chapter 2 identified the field failure modes that have been observed for PV modules. Once failures have been observed, it is up to the module manufacturer to develop improved designs, component materials and processes to manufacture the modules so that they do not suffer these failures. If, however, the only way to assess the impact of a product change is to deploy it in the field and wait to see what happens, improvements to the technology will take a long time. In order to speed up this process, we look for accelerated stress tests (ASTs) that cause the same failures to occur in the modules that have been observed in the field, but of course in much shorter times.

We need to look at each of the failure modes described in Chapter 2 and try to determine what stress or stresses occurring in terrestrial environments caused these failures. Was it?

- Operation at high temperature
- Changes in temperature
- High humidity
- Wind or snow loading
- UV exposure
- Or perhaps a combination of several or all of the above or something else.

Once the driving force for the failure mode has been identified, we can then try to accelerate that stress to cause the failure to occur in the laboratory in a much shorter time period.

In developing ASTs, we must cause degradation. The degradation occurring in the AST must be due to the same failure mechanism we saw outdoors. If not, our accelerated test is accelerating the wrong mechanism and we may learn nothing about how to improve the modules. Because the AST is causing the same failure as seen in the field there is a chance that we can determine an acceleration factor (AST versus field) for this one failure mode remembering that this acceleration factor may be specific to one geographic climate, one mounting system and one module technology. (See Chapter 9.)

It is important to understand why ASTs are so important. Since ASTs duplicate field failure modes they can be used to develop longer life – more reliable modules. Table 3.1 outlines the process one utilizes with AST to improve the module's ability to survive one particular failure mode.

In developing ASTs, it is important to realize that some processes can be accelerated more easily than others. Trying to accelerate 30 years of field performance of PV modules

Photovoltaic Module Reliability, First Edition. John H. Wohlgemuth.
© 2020 John Wiley & Sons Ltd. Published 2020 by John Wiley & Sons Ltd.

Table 3.1 Process flow for using accelerated stress tests (ASTs) to improve module lifetime and reliability.

A module type suffers a particular failure in the field.

Find a specific accelerated stress test (AST) that causes the same failure mode.

Improvements are made to the design, materials or processes of the module in order to alleviate the failure.

The impact of the improvements is then measured by applying the specific AST identified above to the improved module.

If the AST results show substantial improvement, the improved module can be deployed in the field to validate the improved product.

The improved module may now be sold commercially, especially if the AST you used is part of the Qualification Test Sequence (see Chapter 4).

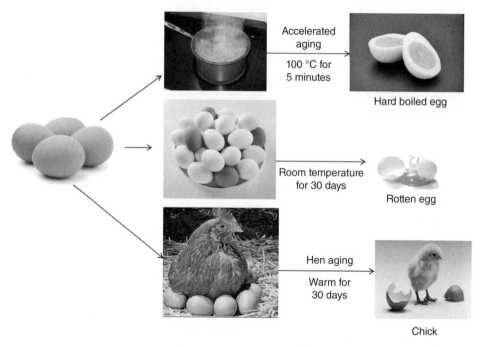

Figure 3.1 Accelerated testing of a chicken egg.

into a 3-month accelerated test program is a little like trying to hatch a chicken in six hours as shown in Figure 3.1. In the field, it takes the hen about 30 days of sitting on the egg to get it to hatch. If the same egg is left at room temperature for 30 days you end up with a rotten egg instead of a chick. On the other hand, if you try to accelerate the process by placing the egg in boiling water for only a few minutes you end up with a hard-boiled egg. Neither of which is the result that you want.

Development of the appropriate ASTs may seem like a daunting task but, in reality, we have over 35 years of experience with applying ASTs to PV modules. Table 3.2 provides a list

Table 3.2 Accelerated stress tests (ASTs) routinely used for PV modules.

Thermal Cycling
Damp Heat
Humidity Freeze
Ultraviolet (UV) Light Exposure
Static Mechanical Load
Cyclic (Dynamic) Mechanical Load
Reverse Bias Hot Spot Test
Bypass Diode Thermal Test
Hail Test

of ASTs that have been found to be useful for testing PV modules. The subsequent sections will describe each of these ASTs indicating what weather and stress phenomena they accelerate and what failure modes they are best at exposing in PV modules. Details of how these tests are usually implemented will be discussed in Chapter 4 on Qualification Testing. However, it is important to remember that in reliability research, experimentation with various stress levels of these tests can provide us with valuable information about the failure modes being observed. Running the tests at multiple-stress levels (say three temperatures or multiple humidity levels) may provide the necessary information to determine the acceleration factor for that failure mode. (See Chapter 9 for more details.)

You may have noticed that there is no listing in Table 3.2 for High Temperature Operation. There can be no doubt that operation at high temperature does reduce the lifetime of most electronic components including PV modules. However, the PV industry ordinarily does not use high temperature alone as an AST. Higher temperatures are used in conjunction with other stresses, for example, UV exposure or humidity, where the higher temperature accelerates the impact of additional stress. This is consistent with what is seen in the field where higher operating temperatures usually accelerate the impact of these other stresses. We really don't see failures due entirely to higher temperature operation unless it is an excursion beyond normal module operating temperatures. One exception is the failure of bypass diodes due to their operating at higher temperatures. In reality, the bypass diode thermal test is really a higher operating temperature test, but restricted to its impact on the bypass diodes not the whole module.

An excellent history of the development of accelerated stress testing for PV modules can be found in the paper by Osterwald and McMahon [1].

3.1 Thermal Cycling or Change in Temperature

A number of the failure modes discussed in Chapter 2 are a result of the stresses caused by thermal expansion and contraction. We all realize that there is a daily temperature cycle from the lowest temperatures at night through the highest temperatures during the middle of the afternoon. Diurnal temperature cycles were recognized as an issue from the earliest days of PV deployment and testing. However, it turns out that stresses on PV modules from the diurnal cycle are, in many cases, not the major cause of thermal stress because the diurnal temperature change takes place relatively slowly.

PV modules can suffer from more rapid temperature change as the sun goes in and out of the clouds. Under full sun exposure ($1000\,W/m^2$) most PV modules increase in temperature by about 30 °C above the ambient temperature of the air in the vicinity of the module. When a cloud suddenly obscures the sun, the module quickly cools down approaching the temperature of the air around it. In partly cloudy environments, PV modules experience these rapid temperature changes resulting in many more temperature cycles than just one per day.

Thermal cycling has been identified with the following failure modes:

- Broken interconnects
- Broken cells
- Electrical bond failures (solder bonds in particular)
- Module open circuits with potential for arcing due to any of the above failures
- Junction box and frame adhesion problems.

Since the earliest days of PV (going back to the use of PV in space applications) thermal cycling has been one of the ASTs applied to modules. The parameters used to set the stress levels in Thermal Cycling include:

- The high and low temperatures of the cycles
- The frequency of the cycles (how many cycles per day)
- The soak level or amount of time at the highest and lowest temperatures
- The maximum (and maybe minimum) allowable rate of temperature change
- Control of the humidity level in the chamber during the thermal cycling
- The total number of cycles
- Whether there is any current flow through the module during the temperature cycling.

Thermal Cycling will be used to evaluate the ability of the module to withstand thermal mismatch, fatigue and other stresses caused by repeated changes of temperature.

3.2 Damp Heat

Humidity has been recognized as a potential cause of degradation in many electronic products including PV modules. Penetration of moisture into the package of an electronic device has the potential to cause corrosion and loss of adhesion between layers. The temperature of the package influences the rate of moisture ingress, the amount of moisture that can be absorbed within the packaging material and the reactivity rate of the moisture. So, most moisture-related degradation mechanisms are thermally activated. To accelerate moisture ingress and the resultant reactions, we use higher temperatures along with high humidity and call the test "Damp Heat."

Damp heat was one of the earliest tests used on solar arrays for space. These tests were not to evaluate the performance of the arrays in space, but rather to verify that they could survive their exposure in Florida before their launch from Kennedy Space Center.

This storage in the humid environment of Florida led to development of improved metallization systems for solar cells. Most space cells switched from Ti/Ag (which is moisture sensitive) to Ti/Pd/Ag grid lines (which are not sensitive to moisture) as a result of the early experiences at the Cape and the subsequent use of damp heat testing for space arrays [2]. The Ti/Pd/Ag cells were then adopted for some of the first terrestrial applications in the 1970s. This was an important step as it meant the earliest terrestrial cells had moisture-resistant cell metallization.

Damp heat has been identified with the following failure modes:

- Corrosion of cell metallization, interconnects and bus bars
- Delamination/loss of adhesion between layers
- Loss of elastomeric properties of encapsulant or backsheet
- Glass corrosion
- Junction box failures
- Adhesive failures
- Potential-Induced Degradation
- Electrochemical corrosion of Transparent Conductive Oxides (TCOs)

The parameters used to set the stress levels in Damp Heat testing include:

- Humidity level at which the test is run
- Temperature level at which the test is run
- The total exposure time
- Whether there is voltage applied to the module during the damp heat test.

Damp Heat will be used to evaluate the ability of the module to withstand the effects of long-term penetration of humidity.

3.3 Humidity Freeze

One contributor to failure is the presence of voids within the lamination package. Once a void is present moisture can condense within the void, particularly when the module cools down at night. This liquid water is more damaging than humidity in the polymeric materials. Liquid water corrodes the metals it comes into contact with and dissolves sodium out of the glass. Some of the worst observed field failures are a direct result of poor adhesion within the module package. One such example was shown in Figure 2.7.

So how do you design an AST for this failure mode? After a variety of attempts to develop a useful moisture test for adhesion within the PV package, Jet Propulsion Laboratories (JPL) introduced the humidity-freeze test in Blocks II-V [3]. (See Chapter 4 for a detailed review of the whole JPL PV Reliability Program.) In the humidity-freeze test, moisture is pumped into the module package using a damp heat exposure similar to what was discussed in Section 3.2. After a significant amount of moisture has been driven into the encapsulant, the whole module is frozen. If there are any voids or delaminations in the module, liquid water will condense in them when the module is cooled. As the cooling continues, this liquid water will freeze. As we all know, ice has a lower density than liquid water so it expands upon freezing. This expansion of ice within voids in the module pushes the package apart. Small voids become large easy-to-see voids and delaminations after the Humidity-Freeze test.

When the Humidity-Freeze test is used on PV modules, it identifies any interfaces that are not well adhered. For EVA-based modules, it also identifies undercured material. It is an excellent test for determining how well all of the interfaces within a PV module have been adhered. The Humidity-Freeze test has been identified with the following failure modes:

- Delamination
- Inadequately cured encapsulants
- Adhesive failures

Several module manufacturers have tried using an enclosed frame design – a box frame. While such frames have good mechanical strength, if improperly designed, they can fill with liquid water and then split apart when the water freezes. The Humidity-Freeze test can also be used to test for this problem.

The parameters used to set the stress levels in Humidity-Freeze testing include:

- The high and low temperatures of the cycles
- The humidity level during the high temperature soak
- The frequency of the cycles (how many per day)
- The soak level or amount of time at both the highest temperature and the lowest temperature
- The maximum (and maybe minimum) allowable rate of temperature change
- Whether there is required control of the humidity level in the chamber while the temperature is changing or during the freeze cycle
- The total number of cycles
- Whether there is any current flowing or voltage applied to the module during the testing.

Humidity Freeze will be used to evaluate the integrity of the PV module package, that is how well it is adhered together.

3.4 Ultraviolet (UV) Light Exposure

UV light is known to degrade some polymers and adhesives. Therefore, use of a UV light exposure should be an important part of any accelerated stress testing program for PV. One of the critical questions for this testing relates to how closely the spectrum of the accelerated UV source can be matched to the short wavelength end of the solar spectrum. Other issues involve selecting a temperature for the exposure as higher temperatures almost always accelerate UV-induced damage.

Another issue with accelerated UV testing is whether a reciprocal relationship exists between the intensity of the UV light and the duration of the exposure. For example, will tripling the UV exposure rate mean that it will only take one-third of the time to achieve the same level of degradation or damage. Another way of looking at this question is whether photochemical reactions depend simply on the total energy absorbed and are independent of intensity and time. For many years, reciprocity was assumed for UV testing.

However, there are a number of references that provide experimental evidence of considerable departure from reciprocity for UV testing in many polymer systems [4, 5].

This means that UV exposure levels must be kept at similar levels to that experienced in the field if you wish to have results that can be compared to actual outdoor exposures. You get some acceleration because, in the laboratory you can apply the UV continuously, while in the field you typically only get four to six equivalent sun hours per day. So continuous exposure may give you an acceleration factor of four to six times, but it is very difficult to reach much higher acceleration rates while ensuring that you are seeing the same degradation mechanisms with the same acceleration rates as you see in the field. Therefore, to simulate long-term UV exposures requires very long test times. Some attempts have been made to accelerate the UV damage using higher test temperatures, but recent work has shown that this approach may not be duplicating the correct failure modes either. This work by the Photovoltaic Module Quality Assurance Task Force (PVQAT) Task Group 5 will be discussed in Chapter 7.

UV exposure has been identified with the following failure modes:

- Encapsulant discoloration
- Delamination/loss of adhesion between layers
- Loss of elastomeric properties of encapsulant or backsheet with subsequent cracking of the backsheet leading to ground faults
- Degradation of the polymers in junction boxes, connectors and cables

The parameters used to set the stress levels in UV testing include:

- Type of light source – usually either xenon arc if you want the whole spectrum or fluorescence if just the UV portion of the spectrum is adequate.
- Spectral distribution of light, which is based on light source and filter used.
- Intensity of the UV light source
- Temperature of the sample
- Humidity of the chamber and sample
- Types of cycle – different cycles are used to look at different degradation modes. Everything from continuous illumination to cycles with UV light and water spray have been used.
- Duration of exposure

The UV Test will be used to identify those materials and adhesive bonds that are susceptible to UV degradation.

3.5 Static Mechanical Load

Modules must be capable of withstanding the mechanical forces they are likely to encounter during their outdoor exposure. They have to be pointed at the sun during operation so must be mounted in either a stationary system with optimum orientation or tracked to follow the sun. In either case, the module must be capable of withstanding wind and snow loads as mounted on the support structure. Dead weights have been utilized to determine whether the module itself can survive the design wind and snow loads with a sufficient safety factor. It is extremely important to ensure that the modules are mounted in exactly the way they will be in the field during the static mechanical load test. If not, the test results will not be indicative of field results (see Figure 2.23). This has often been the case when installers have invented their own module mounting methods, but have not repeated the static load tests with their appropriate geometry.

Static mechanical load has been identified with the following failure modes:

- Structural failures
- Broken glass
- Broken interconnect ribbons
- Broken Cells

The parameters used to set the stress levels in the static mechanical load test include:

- Number of load cycles
- Amount of loads and are they all the same or do you test for wind load first and then a higher snow load?
- Directions of load – just downward or do you also test for wind uplift and are the loads the same or different for the two directions?
- Duration of load

The Static Load Test will be used to evaluate the ability of the module to withstand the static loads it is expected to experience during its deployment.

3.6 Cyclic (Dynamic) Mechanical Load

The static load test discussed above is usually used to evaluate whether the module itself can survive the expected mechanical loads. However, there are many reported conditions under which the modules themselves survived, but one or more of the components did not. Specifically, mechanical stress on the module can result in breakage of cells, interconnect ribbons and solder bonds. One such stress is the oscillating of a module due to variable winds. Another is the shaking that a module may experience while being transported to the installation site or the stress it sees during installation.

There are a variety of ways to simulate such vibrational stress. However, a fairly simple approach that has correlated fairly well with damage observed within the module is the use of a cyclic (dynamic) mechanical load test. This sort of test was originally proposed by JPL and was included in the Block IV/V qualification tests (see Chapter 4). It was probably included in the JPL sequence because it was a staple of their testing for space arrays. The dynamic stresses on a space array during missile launch are surely much larger than the levels experienced during the lifetime of a terrestrial array. However, a less stressful version turned out to be a very useful test to identify cells that had been damaged during processing [6]. This test is now in the process of being added to the Qualification Test sequence (see Chapter 10).

The cyclic (dynamic) mechanical load has been identified with the following failure modes:

- Broken interconnect ribbons
- Broken Cells
- Electrical bond failures

The parameters used to set the stress levels in the cyclic (dynamic) mechanical load test include:

- Number of load cycles
- Shape of load cycles (sign, square wave, etc.)
- Magnitude of loads and are they the same for all directions and cycles
- Duration of load cycle

The Cyclic (Dynamic) Load Test will be used to evaluate if components within the module including solar cells, interconnect ribbons and/or electrical bonds within the module are susceptible to breakage or if edge seals are likely to fail due to the mechanical stresses encountered during installation and operation.

3.7 Reverse Bias Hot Spot Test

As described in Section 2.10, application of a reverse bias on a cell can result in overheating and possibly damage. Therefore, a test is needed to evaluate the susceptibility of a PV module type to permanent damage caused by individual cells being forced into reverse bias operation. The most straightforward way to perform such an evaluation is to determine worst-case shadowing for a particular cell and then to expose the cell to that worst-case shadowing while the remainder of the module is operating under normal conditions.

Details of such a reverse bias hot spot test should include the following steps:

- Determining which cells are most susceptible to being damaged by shadowing. Usually this means finding cells with localized regions of low-shunt resistance which have the most leakage current when the cell is reverse biased.
- Experimentally changing the percentage of the shadowing applied to the most susceptible cells identified above to determine the worst-case shadowing condition. That is what shadow condition results in the most heat generation.
- Deciding how long the worst-case shadowing should be continued in order to achieve the highest temperature and, therefore, to identify potential reverse bias hot spot issues. Does operation at this temperature result in damage to the cells, solder bonds, encapsulant, or even backsheet?

The Reverse Bias Hot Spot Test will be used to evaluate the ability of the module to withstand reverse bias hot-spot heating effects, e.g. solder melting or deterioration of the encapsulation, caused by shunted, mismatched or broken cells, shadowing or soiling.

3.8 Bypass Diode Thermal Test

As stated in Section 2.10, bypass diodes are utilized to limit the reverse bias voltage on shadowed or otherwise current limited cells. The last chapter discussed a test for reverse bias hot spots which can be used to determine whether the cell and bypass diode circuit design is adequate to protect the cells from reverse bias hot spot damage. However, that test says nothing about the ability of the bypass diodes themselves to survive long-term operation in the module.

Diode manufacturers provide a rating for their diodes that specifies the maximum junction temperature rating for continuous operation of that diode. Diodes should be able to operate continuously (virtually forever) as long as the operating temperature remains below this maximum junction temperature rating. However, if because of the module design and diode selection, the diode operates above its maximum junction temperature rating; its lifetime will be limited. So, a thermal test for the bypass diode should be designed to determine what temperature the diode junction will operate at under normal operating conditions, when the diode is conducting current around shadowed cells or a failed cell string.

The parameters used to set the stress levels in the bypass diode thermal test should include:

- How to measure the diode junction temperature.
- Selection of an ambient temperature for the module used during the test.
- Whether all the diodes in the module should be in the "on" or conducting state or just the one under test.
- Amount of current flow through the diode(s).
- Length of test – do you test for a specific amount of time or until thermal equilibrium is reached?

The Bypass Diode Thermal Test will be used to evaluate the adequacy of the thermal design and relative long-term reliability of the bypass diodes used to limit the detrimental effects of module hot-spots.

3.9 Hail Test

In the early days of PV, many commercial modules did not have glass superstrates. Some of these modules suffered damage from hail storms. This was observed during the JPL program that will be discussed in Chapter 4. To prevent manufacturers from designing and building modules that could not survive the expected level of hail stress, JPL developed a hail test [7]. Details of that test will be given in Chapter 4. This section will discuss the overall design of the test and provide the parameters that must be selected to run such a test.

The first question that arose was whether the test article had to be an ice ball or if other projectiles like steel balls could be used. It didn't take many experiments to verify that there was huge difference in results depending upon whether you used an ice ball, which broke up upon impact, or a metal ball that just bounced off of the module. Balls other than ice tended to cause much more damage to the module, so since the early 1980s, hail tests have used real ice balls.

The parameters used to set the stress levels in the hail test include:

- How to make and inspect the ice balls.
- Size of the ice balls.
- Temperature of the ice balls.

- Velocity of the ice balls.
- Temperature of the module.
- Locations on the module that should be impacted by the ice balls.

The Hail Test will be used to verify that the module is capable of withstanding the exposure to hail.

References

1 Osterwald, C.R. and McMahon, T.J. (2009). History of accelerated and qualification testing of terrestrial photovoltaic modules: a literature review. *Progress in Photovoltaics: Research and Applications* 17: 11–33.

2 Rauschenbach, H.S. (1976). *Solar Cell Array Design Handbook*, Volume 1. NASA-CR-149364 https://ntrs.nasa.gov/search.jsp?R=19770007250 (Accessed 28 August 2019).

3 Smokler, M.I., Otth, D.H., and Ross, R.G. (1985). The Block Program Approach to Photovoltaic Module Development. 18th IEEE PVSC, Nevada, USA (21–25 October 1985).

4 Jorgenson, G.J., Bingham, C., King, D.E. et al. (2002). Use of uniformly distributed concentrated sunlight for highly accelerated testing of coatings. In: *ACS Symposium Series 805* (eds. J.W. Martin and D.R. Bauer). Washington D.C.: American Chemical Society.

5 Scott, K. and Hardcastle, H. III (2009). A new approach to characterizing weathering reciprocity in Xenon Arc weathering devices. In: *Service Life Prediction of Polymeric Materials*, vol. 83 (eds. J.W. Martin, R.A. Ryntz, J. Chin and R.A. Dickie). Boston: Springer.

6 Wohlgemuth, J.H., Cunningham, D.W., Placer, N.V. et al. (2008). The Effect of Cell Thickness on Module Reliability. 33rd IEEE PVSC in California, USA (11–16 May 2008).

7 Moore, D., Wilson, A., and Ross, R. (1978). Simulated Hail Impact Testing of Photovoltaic Solar Panels. *Proceedings of 24th Annual Technical Meeting, Institute of Environmental Sciences*, Texas, USA (18–20 April 1978).

4

Qualification Testing

Chapter 3 described the types of accelerated stress tests that are typically applied to photovoltaic (PV) modules. These tests are used in research to evaluate how well modules perform in relation to the specific stresses applied during these tests. However, what the industry needs is a defined set of accelerated stress tests that can be applied to all modules in the same way. Such a set of tests are called Qualification Tests or often in Europe are referred to as Type Approval Testing.

Qualification tests are a set of well-defined accelerated stress tests developed out of a reliability program. They incorporate strict pass/fail criteria. Hoffman and Ross [1] defined the purpose of qualification testing as a means of rapidly detecting the presence of known failure or degradation modes in the intended environment(s). The stress levels and durations are limited so the tests can be completed within a reasonable amount of time and cost. One of the goals of Qualification testing is for a significant number of commercial module types to pass and that all subsequent production modules will be built the same way as the modules that were tested. Passing the Qualification test says the product meets the specific set of criteria, but doesn't predict product lifetime nor indicate which product will last longer or which will degrade in operation. However, if properly designed, the Qualification test will be a good indicator that modules passing the test sequence will not suffer from infant mortality – that is they will survive for a reasonable amount of time in the field. The real usefulness of such a Qualification test sequence can only be validated by assessing the field performance of products that have successfully passed the test sequence.

The initial development of a Qualification test sequence for PV modules occurred during the Jet Propulsion Laboratory's (JPL) Block Buy Program in the late 1970s and early 1980s. Much of our present testing protocol is based on the JPL efforts, so this chapter will begin with a description of the JPL program and discussions of the lessons learned during that program. The second section will explain how PV-module Qualification testing transitioned from JPL to IEEE to Solar Energy Research Institute (SERI) and finally to IEC, where it remains today. The third section will describe the details of the latest version of the IEC Qualification Test Sequence, IEC 61215 [2]. The fourth section discusses how use of the Qualification tests have improved the reliability of PV modules. The fifth section in this chapter discusses the limitations of Qualification testing and identifies some of the additional steps that are being taken to further improve module reliability and lifetime. Finally, the last section provides the details of the IEC Module Safety Tests (MST).

Photovoltaic Module Reliability, First Edition. John H. Wohlgemuth.
© 2020 John Wiley & Sons Ltd. Published 2020 by John Wiley & Sons Ltd.

4.1 JPL Block Buy Program

In 1975, the US Government initiated research on PV with the goal of developing it as a viable alternative energy option. The Flat-Plate Solar Array Project (FSA) at the JPL was a part of this effort designed to improve the performance of PV modules. One part of FSA program was the JPL Block Buy program, in which PV modules were procured and fielded to evaluate performance and reliability. JPL prepared design and test specifications and a statement of work through which they procured the modules. These specifications evolved over time in order to stimulate the proposers to take maximum advantage of the latest PV technologies. See Ref. [3] for one example of these specifications.

For each phase of the program (Block I to Block V) multiple contracts were issued to different manufacturers for designing and building modules. Each contractor was encouraged to continue work on their module design until successfully passing the Qualification tests or abandoning the proposed design. Over the course of the program. both the module design and the required Qualification tests changed dramatically. Table 4.1 provides a comparison of the Qualification tests used in each Block. Table 4.2 provides a brief comparison of the module designs for each Block. An excellent article summarizing the entire program can be found in Ref. [4]. The following subsections provide a summary of each of the Block Programs.

Block I (1975) – In Block I, JPL procured commercially available modules from four contractors – Sensor Technology, Solarex Corporation, Solar Power Corporation and Spectrolab. This procurement was designed to establish the state-of-the-art in terrestrial PV at that time. The procured modules were all small, single-crystal Silicon battery charging modules with a single series string of cells, single interconnect ribbons between cells and in most, if not all, cases only one solder bond per side on each cell. All four

Table 4.1 Qualification tests from Jet Propulsion Laboratory (JPL) block buys.

Test	Block I	Block II	Block III	Block IV	Block V
Thermal Cycles −40 to +90 °C	100	50	50	50	200
Humidity Test	70 °C, 90% RH 68 hrs	5 cycles 40 to − 23 °C, 90% RH	5 cycles 40 to − 23 °C, 90% RH	5 cycles 40 to − 23 °C, 90% RH	10 cycles 85 to − 40 °C 85% RH
Hot Spot Test					3 cells for 100 hrs
Mechanical Load Test		100 cycles ± 2400 Pa	100 cycles ± 2400 Pa	10 000 cycles ± 2400 Pa	10 000 cycles ± 2400 Pa
Hail Test				9 impact ¾″ – 45 mph	10 impacts 1″ – 52 mph
Hi Pot Test		< 15 μA at 1500 V	< 50 μA at 1500 V	< 50 μA at 1500 V	< 50 μA at 2*Vs + 1000 V

Table 4.2 Properties of Jet Propulsion Laboratory (JPL) block buy modules.

Property	Block 1	Block II	Block III	Block IV	Block V
Power Range (W)	5 to 13	11 to 34	11 to 35	19 to 85	70 to 185
Module Efficiency	4.8% to 6.5%	6% to 7.4%	6.5% to 8.4%	9.3% to 13.6%	9.4% to 13.3%
Wafers/Cells	All CZ	All CZ	All CZ	6 CZ 2 multi	3 CZ 1 multi 1 EFG ribbon
Construction	1 Glass superstrate 3 substrate designs	1 Glass superstrate 3 substrate designs	2 Glass superstrate 3 substrate designs	All Glass superstrates	All Glass superstrates
Encapsulant	Cast Silicone	3 Cast Silicone 1 PVB	4 Cast Silicone 1 PVB	1 cast Silicone 4 PVB 3 EVA	All Laminated EVA
Interconnect Redundancy	None	Minor	Minor	Much	Much
Bypass Diodes	None	None	None	Yes	Yes

module types used cast silicone as the encapsulant, while three out of four used a substrate design with one having a glass superstrate.

The Qualification test sequence for Block I was very simple, requiring only 100 thermal cycles and a 68-hour damp heat test as shown in Table 4.1. It turned out that all four contractors had problems passing 100 thermal cycles. Looking back today, that is not surprising, since these modules contain many of the issues identified in Sections 2.1 and 2.6 as leading to failures of interconnect ribbons and solder bonds. If these modules had not used a small number (between 18 and 25) of small cells (the largest being 87 mm in diameter) they would probably never have been able to survive 100 thermal cycles.

Modules delivered for Block I were deployed in the field, as well as being utilized in the development of additional accelerated stress tests, for incorporation in the Qualification sequences in subsequent Blocks.

Block II (1976) – For Block II, JPL provided a module specification for the manufacturers to meet. The specifications were designed to guide the manufacturers to design and build modules that could ultimately lead to meeting the Department of Energy (DOE) goals for PV in terms of cost, efficiency and reliability. Once again, JPL selected the same four contractors – Sensor Technology, Solarex Corporation, Solar Power Corporation and Spectrolab. The procured modules were still all single-crystal Si modules with a single series string of cells and single interconnect ribbons between cells, but many of the Block II modules did have redundant solder bonds. Three of the four module types used, cast silicone as the encapsulant, while the fourth used Polyvinyl butyral (PVB). As in Block I, three out of four module types used a substrate design with one having a glass superstrate.

For the first time, the modules were required to pass a more complete qualification test sequence that included thermal cycling, humidity freeze cycling, mechanical load cycling and a hi-pot test as shown in Table 4.1. It appears that the trouble contractors had passing the 100 thermal cycles in Block I led JPL to reduce the thermal cycle requirement to 50 cycles in Block II. In hindsight, this was not a good decision but that was later corrected in Block V. Finally, a Twist Test was added in Block II. This was not really an accelerated stress test, but was designed to ensure that the modules could flex enough to fit into JPL's mounting system. However, this Twist Test survived through Block V and made it into the first editions of the IEC Qualification standards, IEC 61215 and IEC 61646. JPL was planning to drop the Twist test in Block VI. IEC finally did drop it from the second editions of IEC 61215 and IEC 61646.

Modules delivered to JPL for Block II were deployed at 16 sites providing a broad range of environments: mountain, desert, marine, hot and dry, hot and humid, cold, moderate, windy, and high pollution [5]. Modules at these sites were evaluated continually and the results fed back into the program. A summary of the Block II results can be found in the report by Smokler [6].

Block III (1977) – The Block III modules were production quantities (tens of kilowatts) of the Block II modules, with slightly revised specifications. JPL selected five contractors – Arco Solar, Motorola, Sensor Technology, Solarex Corporation, and Solar Power Corporation. The procured modules were still all cry-Si modules. All but the Motorola modules still had a single-series string of cells with a single interconnect ribbon between cells, but many of the Block III modules did have redundant solder bonds. Four of the five module types used cast silicone as the encapsulant, while the fourth used PVB. In this case, three out of five module types continued to use a substrate design with two having glass superstrates.

The Block III Qualification test was identical to that for Block II except for the limits on hi-pot testing leakage current. The allowable level actually increased from 15 to 50 μA measured at 1500 V. As modules were getting larger, this was a necessity and, as will be seen later, the IEC Qualification Test Sequence finally made the leakage current a function of the area of the module under test.

The Block III modules were also deployed at the 16 sites used for Block II. In addition, Block III modules were provided for a number of larger systems both in the US and around the world. A summary of the Block III results can be found in the report by Smokler [7].

Some Lessons Learned from Blocks II and III – The deployment of modules from Blocks II and III provided some useful information about failure modes that were not adequately addressed in the Block II and III Qualification Test Sequence.

Hail did significant damage to modules built without tempered glass superstrates. Modules suffered from broken cells and broken annealed glass superstrates. In response to this, JPL added a Hail Test to the Qualification Test Sequence in Block IV, while many cry-Si module manufacturers began using tempered glass superstrates.

Some Block II and III modules were used in desert environments including:

- Pagago Indian Reservation in Arizona
- Tanguze, Upper Volta
- Natural Bridges, Utah.

Some of the modules began failing in the desert after three to five years due to broken interconnects, failed solder bonds and/or broken cells that resulted in total loss of module power. In response to these failures, module manufacturers started building in redundant interconnects and stress relief. Most new module types used glass superstrate construction, reducing the thermal expansion and contraction. Finally, in Block V JPL increased the number of Thermal Cycles to 200 to better evaluate module performance versus thermal cycling in the field.

Block II and III modules were used in what, at that time, was considered a large (60 kW), high voltage (600 V) system in Mt. Laguna, California. Part of that array was built using Solar Power modules (40–4″ diameter single-crystal cells in series) with no bypass diodes in the module although the system designer did add one bypass diode for every module. These Solar Power modules began suffering from reverse bias hot-spot failures – that is, when shadowed, they had a tendency to overheat and catch fire. In response to this, JPL added a Hot Spot Test to the Block V Qualification Test Sequence. Module manufacturers began to add, or at least offer as an option, a bypass diode approximately every 18–24 cells in most cry-Si power modules.

Block IV (1980) – The Block IV Specification contained more rigorous requirements than the previous blocks. It was designed to take advantage of the research that had been going on in JPL's FSA Project [8]. Proposers to Block IV had to demonstrate why they thought the module types they were proposing could meet the cost, reliability, and lifetime goals of the DOE PV program. The modules proposed and built for Block IV took advantage of many of the newest materials and designs recently developed. Block IV also provided an opportunity for manufacturers to propose modules for use in intermediate load or residential applications with different sets of requirements for each.

JPL selected six contractors to design and fabricate modules for intermediate load applications – Arco Solar, ASEC, Motorola, Photowatt, Solarex, and Spire. They also selected two contractors to design and fabricate modules for residential applications – General Electric (GE) and Solarex. While all of the proposed modules were still cry-Si, they were no longer all single-crystal silicon as both types of Solarex modules used cast multi-crystalline silicon cells. All of the Block IV modules had redundant interconnect ribbons and multiple solder bonds per cell side. Five out of the eight module types had some paralleling of circuits with four of those having cross ties. Three of the designs had more than one bypass diode per module.

Block IV saw major changes in the PV packaging. All of the module designs had glass superstrates. Only one design, the GE roof shingle, continued to use a cast silicone encapsulant. Four of the module types used PVB, while three used the newly developed Ethylene Vinyl Acetate (EVA). The Block IV modules look more like today's modules than the small battery charging modules that were part of Block I.

There were really only two significant changes to the Qualification tests between Blocks III and IV. One of these was the addition of the hail test, which drove all of the module designs to use tempered glass superstrates. The second change was the increase in the number of cycles for the mechanical load test from 100 to 10 000. It is unclear why this change was made, but long term this was a poor decision as performing 10 000 cycles with a 2400 Pa load could severely restrict the size and design of modules in the future. Most of today's commercial modules could not survive this test although they survive the dynamic mechanical stresses they see in the field. This test did remain in some of the later Qualification tests; but it was never included in the IEC Qualification Test Sequence.

A summary of the Block IV results can be found in the report by Smokler [9].

Encapsulant Lesson Learned Between Blocks IV and V – The initial transition for PV encapsulants was from cast silicone to PVB. The first non-silicone modules in the Block Buy Program were the Spectrolab modules in Block II and the Arco Solar modules in Block III. By Block IV, there were more types with PVB encapsulant than any other material. Indeed, during this timeframe Arco Solar, the largest producers of PV modules in the world at the time, had switched most if not all of their product lines to a PVB encapsulant. It did not take long for some of the PVB modules, especially those in high-voltage systems, to start showing signs of corrosion of cell metallization as shown in Figure 2.9. Work at JPL [10] demonstrated that the conductivity of PVB reduced significantly when exposed to moisture resulting in high-leakage currents, migration of the silver metallization and severe power loss. Basically, the PV industry abandoned PVB and all of the Block V modules used EVA encapsulants.

Block V (1981) – The Block V Specification was more rigorous than those from the previous Blocks. Once again, there were actually two specifications, one for Intermediate Load Applications and one for Residential Applications. JPL selected five contractors for Block V, three for residential modules (GE, Mobil Solar Energy Corporation and Spire Corporation) and two for intermediate load applications (Arco Solar and Solarex). All five used cry-Si cells, but only three (Arco Solar, GE, and Spire) used single crystal cells. Solarex continued its development using semi-crystalline Si cells, while Mobil Solar used its Edge Defined

Film Growth (EFG) cells. All of these modules had parallel circuitry and most had cross-ties although this varied considerably from module type to module type. Once again all of these module types had built in bypass diodes although several only had one diode per module. Because of the paralleling, the worst-case situation was one diode per 36 cell series block, which would probably not be considered safe today, but it appears that this design did pass the new Hot Spot Test.

As previously stated, all of the Block V modules had glass superstrates and used laminated EVA as the encapsulant. With laminated EVA, a backsheet is required. In Block V, a number of module manufacturers began using backsheets with layers of Tedlar, polyester, and Tedlar. Several added aluminum foil to form vapor barriers. Some of these materials are still used in modules today.

The Block V Qualification Test Sequence had a number of very important changes that proved invaluable in identifying and ultimately helping to eliminate several critical failure modes. The important changes were:

- Increasing the number of thermal cycles from 50 to 200.
- Making the humidity test a true humidity-freeze test where humidity is pumped into the module for 20 hours at 85 °C and 85% Relative Humidity and then it is frozen at − 40 °C.
- Addition of a Reverse Bias Hot Spot Test to aid module manufacturers in determining how to provide adequate bypass diode protection.
- Making the hi-pot test dependent on the highest systems voltage that a module is rated for.

This set of tests has formed the core of the PV Qualification Test Sequence for more than 35 years.

Block VI – JPL was in the planning stages for Bock VI when the whole LSA program was terminated because of the Regan budget cuts to the Renewable Energy Program. There were several important tests under review for incorporation into the Block VI Qualification Test Sequence that ultimately had an impact on the IEC Qualification Test Sequence. Three specific tests under evaluation for addition to the Block VI specification were:

- Ultraviolet exposure test – As we know, all PV modules are exposed to the UV portion of the solar spectrum and many polymeric materials are sensitive to long-term exposure to UV. A UV test has been included in the IEC Qualification Test Sequence.
- Electrochemical corrosion – JPL was considering adding electrical stresses to the Block V Humidity-Freeze test procedure to assess module sensitivity to electrochemical degradation [4]. As we will see in Section 4.3, IEC added a damp-heat test without application of an electric field instead.
- Design criteria and test procedure for bypass diodes – It was recognized that the major failure mechanisms for bypass diodes were temperature dependent. JPL was planning to establish design criteria addressing the adequacy of diode heat sinking by limiting diode junction temperatures to: (i) the manufacturer's maximum allowable operating temperature under worst-case conditions; and (ii) a de-rated temperature for prolonged periods of high operating temperature. IEC ultimately added a Bypass Diode Thermal Test, but not until the second edition of IEC 61215.

Quality Assurance System – In addition to product testing, JPL also realized that if module manufacturers did not have adequate Quality Assurance Systems there would be no guarantee that the production modules would look or operate anything like the engineering samples that successfully passed the Qualification Tests. JPL Quality Assurance (QA) personnel worked with the Block contractors to develop inspection criteria, train personnel and prepare QA plans. These plans were required to show the role of QA in the production process, including inspection criteria for the modules, and provided for JPL review and approval of the QA operations. For large production orders, JPL inspectors were in residence at the contractor sites and performed acceptance inspections there.

The fact that small PV manufacturing companies were introduced to QA systems early in their development was important for PV. While the individual companies may not have used the whole Block Buy QA system for their commercial production of modules, their familiarity with such a system allowed them to see the advantages of having a robust QA system. So, like the Qualification Test, QA systems became part of the operation of most PV module manufacturers early in their history. Chapter 6 will discuss the evolution of Quality Management Systems in PV module manufacturing.

Massive paralleling of Modules – One of the conclusions of JPL's reliability work was that PV modules would be much more reliable if they had parallel circuits with cross ties rather than just series strings of cells because of added redundancy [11]. This was especially true in the early days of PV when there was little or no redundancy of interconnect ribbons or solder bonds and when failures of solder bonds and interconnect ribbons were fairly common. So, as we saw by Block V, all of the module designs included parallel circuits and at least some had cross ties. For a time, modules like this were used in a number of larger systems like the initial SMUD Ranco Seco utility scale project.

The problem with massive paralleling is the trade-off between voltage and current. Circuits have the greatest resistive loss in the lowest voltage/highest current configuration. For example, in one of the Arco Solar arrays at Ranco Seco, the modules had short circuit currents of 25–30 A while the Solarex modules had short circuit currents of 30–35 A. These are huge currents requiring large current carrying capacity wires and extremely robust connectors. Degradation of the contacts within these systems resulted in significant power loss. As the PV industry designed better cell/module combinations with redundant tab across interconnect ribbons, and improved the interconnect materials and soldering processes, series connected modules became much more reliable. Paralleling of cry-Si modules grew out of favor as everyone wanted higher voltage systems to reduce resistive loss in the PV array and to allow for use of more efficient inverters.

Summary of Block Buy Achievements – The JPL Block Buy Program took PV from building small modules for remote site markets to the capability of building larger power modules. Figure 4.1 shows examples of one module type from each of the Blocks. The transition was really dramatic. The Block V products had properties similar to, and looked fairly similar to, modules being fabricated today.

The JPL Block V Qualification Test Sequence became the de facto Qualification Test that modules needed to pass. As will be discussed in the next section, this test sequence served as the basis for the Qualification Tests that followed. Section 4.4 will discuss how successful the Block V Qualification Test Sequence was.

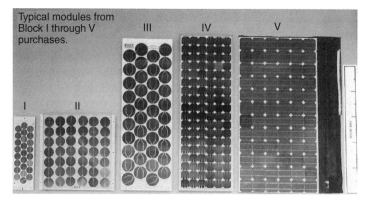

Figure 4.1 Picture of representative samples of Module from each of the Jet Propulsion Laboratory (JPL) block buys [4]. *Source:* reproduced from IEEE PVSC with permission of Ron Ross of the Jet Propulsion Laboratory.

4.2 Evolution of IEC 61215 Qualification Test Sequence

With the end of the Block Buy Program, JPL stopped working on module qualification testing. However, those customers who realized the value of qualification testing for PV modules, continued to ask for product that was "Qualified to Block V." In JPL's view, only the five module types qualified during the Block V program were truly "Qualified to Block V." What customers wanted were modules tested to the test sequence given in Block V as shown in the Block V column of Table 4.1.Without a consensus standard to use, module manufacturers didn't really know what to test to, test labs didn't really know what tests to perform and customers weren't really sure what they were getting when a manufacturer claimed that a module type was "Qualified to Block V." To remedy this situation, several standards' organizations began work on a PV module Qualification Standard. A summary of the results of these efforts is shown in Table 4.3. The following sections will describe how these different standards helped lead to the IEC 61215 Qualification Standard we have today.

EU Specifications 501, 502, and 503 – Through the European Solar Test Installation (ESTI) the European Community started working on PV Qualification Standards at about the same time that JPL was working on Block V. European Standards 501 and 502 had many similarities to the Block V document with the addition of a UV Exposure Test and an Outdoor Exposure Test and a change in the thermal cycle maximum temperature from 90 to 85 °C. This temperature difference is very typical of the differences between European and US PV. Typical module temperatures in the US southwest are much higher than anything normally experienced in Europe.

EU 503 was based on a draft of the first edition of IEC 61215. It was utilized to begin testing to the new standard before it had completed voting by the IEC National Committees. This way, test laboratories had a chance to perform the test sequence and provide feedback on any issues they encountered before the IEC document was finalized.

SERI Interim Qualification Test (IQT) – SERI was working on new thin-film modules, mostly a-Si at the time. The "interim standard" or IQT for these modules was based primarily on Block V with some specific differences addressing the thin-film modules as well as

Table 4.3 History of photovoltaic (PV) module qualification tests.

Title	Year	Technologies Covered
JPL Block V	1981	Crystalline Silicon
EU Specs 501, 502 & 503	1981 to 1991	Crystalline Silicon
SERI IQT	1990	Thin Film a-Si
IEC 61215	1993	Crystalline Silicon
IEEE 1262	1995	All
IEC 61646	1996	Thin Film a-Si
IEC 61215 Ed 2	2005	Crystalline Silicon
IEC 61646 Ed 2	2007	Thin Films (All Types)
IEC 61215-1	2016	Part 1: General Requirements
IEC 6125-1-1	2016	Part 1-1: Specific Requirements for Cry-Si
IEC 61215-1-2	2016	Part 1-2: Specific Requirements for CdTe
IEC 61215-1-3	2016	Part 1-3: Specific Requirements for a-Si
IEC 61215-1-4	2016	Part 1-4: Specific Requirements for CIGS
IEC 61215-2	2016	Part 2: Test Requirements

addition of several safety tests and one test that had been planned for Block VI. The biggest new reliability issue with the a-Si modules of that time was the high-leakage current resulting from inadequate edge isolation of the Transparent Conductive Oxide (TCO). The IQT added a Wet Leakage Current Test to screen for this problem. As we will see later, the Wet Leakage Current Test is now included in the latest version of IEC 61215. The IQT also added a Ground Continuity Test and a Cut Susceptibility Test from UL 1703 [12]. Finally, the IQT incorporated the Bypass Diode Test from Block VI, a version of which is also now included in IEC 61215.

IEC 61215: 1993 – The first edition of IEC 61215 was the first international PV module Qualification Standard. It tried to include all of the best ideas from around the world. It was based on Block V with the replacement of the dynamic mechanical load test because this was judged unsuitable for testing large modules with a static load test. The UV Test and the Outdoor Exposure Test were added from EU Specification 502 although there was actually no UV test procedure given in the first edition. The procedure for UV Testing was not added until the second edition. Finally, a Damp Heat Test was added in response to the corrosion problems that were being seen in the field. A simplified outline of the IEC 61215 first edition test sequence is shown in Figure 4.2.

IEC 61215 incorporated strict pass/fail criteria. The modules were evaluated through visual inspection, peak power measurement and an insulation test before and after each of the accelerated stress tests. The document defined the major visual defects that were grounds for failure. Each module could lose no more than 5% of its peak power after each individual test and no more than 8% through an entire sequence of tests. Finally, each module had to pass the insulation resistance test before and after each of the accelerated stress tests.

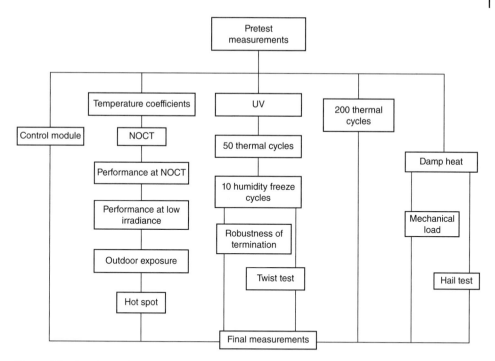

Figure 4.2 Accelerated stress test sequence from the first edition of IEC 61215. *Source:* redrawn by the Author for clarity.

The first edition of IEC 61215 required eight modules which were used in the following ways:

- One control module that was stored in the dark and used to check the repeatability of the solar simulator.
- One module that went through the characterization sequence consisting of:
  - Measurement of the module temperature coefficients of short circuit current (α) and of open circuit voltage (β).
  - Measurement of the Nominal Operating Cell Temperature (NOCT).
  - Measurement of the module performance (I-V curve) at NOCT.
  - Measurement of the module performance at low irradiance.
  - Outdoor Exposure Test
  - Hot Spot Test
- Two modules went through the combined UV/TC/HF test sequence consisting of:
  - UV Test
  - 50 Thermal Cycles
  - 10 Humidity Freeze Cycles.
  Then, one of the two modules went through the Robustness of Termination test while the other went through the Twist Test.
- Two modules went through 200 Thermal Cycles.
- Two modules went through the Damp Heat Test. Then, one of the two went through a static Mechanical Load Test while the other went through the Hail Test.

IEC 61215 soon became the qualification test to pass in order to participate in the PV marketplace, especially in Europe.

IEC 61646: 1996 – The first edition of IEC 61646 was written for the thin-film modules available in 1996 – which were mostly a-Si. The basis for IEC 61646 was the first edition of IEC 61215 with the added thin-film requirements of the SERI IQT particularly the addition of the wet leakage current test. The other changes from IEC 61215 were based on the light instabilities of a-Si modules. The first edition of IEC 61646 had the following difference from the first edition of IEC 61215:

- Addition of a wet high pot test on each module at the beginning and end of each of the test sequences and within a specified time period after the damp heat test.
- Use of thermal annealing and light soak in an attempt to characterize the power loss caused by the different accelerated tests. Modules were light soaked before the measurement leg and before the UV Test. Before each of the chamber tests they were annealed. Finally, they were all light soaked to stabilization before the final power measurements.
- In addition to the 5% pass/fail for maximum power measured after each test as in IEC 61215, there was a requirement that after the final light-soaking in the first edition of IEC 61646, the maximum output power at standard test conditions (STC) should not be less than 90% of the minimum value specified by the manufacturer.

Ultimately, the first edition of IEC 61646 was used to qualify a-Si, CdTe, and CuInGaAs modules.

IEC 61215: 2005 – The second edition of IEC 61215 added some of the tests from IEC 61646 as well as more of the original tests proposed in Block VI and the SERI IQT. The major changes between the first and second Edition of IEC 61215 were:

- Addition of Wet Leakage Current Test (as in IEC 61646) at the maximum-rated systems voltage with the pass/fail criteria a function of the module area.
- Addition of the Bypass Diode Thermal Test.
- Removal of the Twist Test.
- Addition of a Pretest Light Soak of 5kWh/m^2 to stabilize or remove the Light Induced Degradation (LID) from the crystalline silicon module pretest power measurements.
- Making the pass/fail criteria for the insulation test a function of the module area.
- Added the measurement of the temperature coefficient of power (δ).
- Added a procedure for measuring the temperature coefficients of the module using either natural or simulated sunlight.
- Providing a procedure for the UV Test.
- Changed the Mechanical Load Test from two to three cycles.
- Required module maximum power current flow during the 200 thermal cycle test. This was added in response to field failures related to solder bonds that are too small, but which passed the thermal cycle test if there is no current flow [13].

The second edition of IEC 61215 served as the module qualification test for crystalline silicon modules for more than 10 years.

IEC 61646: 2007 – The second edition of IEC 61646 was an updated version based on the second edition of IEC 61215 and designed for the multiple thin-film technologies available at that time (a-SI, CdTe, and CIGS). The major changes between the first and second Edition of IEC 61646 were:

- No longer requires meeting a plus/minus criterion before and after each test, but rather on being within 10% of the rated power after all tests have been completed and the modules have been light-soaked.
- Removed the "Twist Test" as was done for the second edition of IEC 61215.
- Made the pass/fail criteria for Insulation Resistance and Wet Leakage Current tests dependent on the module area.
- Added the measurement of the temperature coefficient of power (δ).
- Added a procedure for measuring the temperature coefficients of the module using either natural or simulated sunlight.
- Deleted the reference plate method for measuring NOCT.
- Revised the Hot-Spot Test.
- Eliminated the edge dip method for performing the Wet Leakage Current Test.
- Changed the Mechanical Load Test from two to three cycles to be consistent with IEC 61215 second edition.
- Added the Bypass Diode Thermal test.

IEC 61215 Series: 2016 – The biggest change in the 2016 version of IEC 61215 was that it covered all PV technologies, replacing earlier versions of both IEC 61215 and IEC 61646. To be consistent with other IEC Qualification Tests, there was also a change in structure so that Part 1 lists general requirements, Part 1-x provides specific requirements for different PV technologies and Part 2 defines the testing methods. In addition to these major structural changes, the following changes were also made in this edition:

- Added a requirement for the initial module power measurements to meet the manufacturer's name plate rating.
- Revised the hot-spot endurance test
- Removed the method for measuring temperature coefficients and referenced IEC 60891.
- Replaced NOCT measured at open-circuit voltage with Nominal Module Operating Temperature (NMOT) measured at peak power.
- Modified the robustness of termination test to include evaluation of both output cables and junction boxes.
- Added a bypass diode functionality test to validate that the bypass diodes are still working after the bypass diode thermal test.
- Added a stabilization procedure for the PV modules, replacing either the light-soaking procedure from IEC 61646 or preconditioning from IEC 61215.

Each commercialized PV technology now has a subpart of IEC 61215-1 that provides guidance on specific test requirements for that technology. So far, those technology specific requirements have been limited to providing:

- The maximum allowable value of reproducibility.
- The current flow to be used during the UV Test and Thermal Cycle Test.
- The stabilization procedure.

- Special instructions for the Hot Spot Test particularly related to whether the module is wafer based or monolithically integrated (MLI).
- Special allowance to perform certain tests under slightly different bias conditions because of stability issues.

Details of the IEC 61215: 2016 Qualification Test Sequence are given in the next section.

4.3 IEC 61215 Test Protocol

A summary of the test sequence from the 2016 version of IEC 61215 is shown in Figure 4.3. This and the subsequent descriptions of the test sequence are the author's interpretation of the IEC 61215 test sequence designed to familiarize the readers with the tests, not to provide alternative methods for performing the tests. Anyone actually performing the IEC 61215 qualification test sequence should use the procedures as published in the IEC 61215 documents.

In the 2016 version of IEC 61215, all of the tests have been given module quality test (MQT) numbers so these will be used in the explanation of module flow and then in the

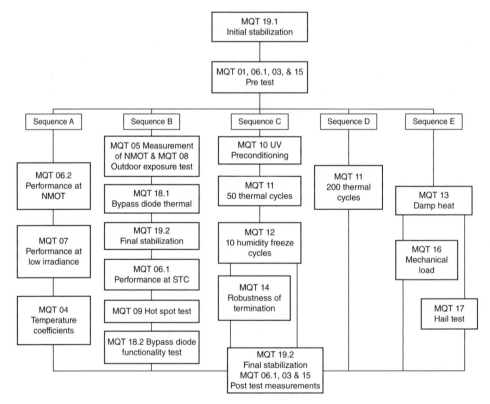

Figure 4.3 Summary of accelerated stress test sequence from IEC 61215: 2016. *Source:* redrawn by the Author for clarity.

descriptions of the test procedures. This edition of IEC 61215 requires ten modules which are used in the following ways:

- All ten modules go through the Initial Stabilization (MQT 19.1) and the pretest measurements which include Visual Inspection (MQT 01), Performance at STC (MQT 06.2), the Insulation Test (MQT 03) and the Wet Leakage Current Test (MQT 15).
- Three modules go through Sequence A which includes measurement of Performance at NMOT (MQT 06.2), Performance at Low Irradiance (MQT 07), and Temperature Coefficients (MQT 04), with one of the three modules then used as a control for the other five sequences.
- One module goes through Sequence B which includes Measurement of MNOT(MQT 05) and the Outdoor Exposure Test (MQT 08); which can be performed at the same time. Then this module goes through the Bypass Diode Thermal Test (MQT 18.1), Final Stabilization (MQT 19.2), measurement of Performance at STC (MQT 06.2), the Hot Spot Test (MQT 09) and the Bypass Diode Functionality Test (MQT 18.2).
- Two modules go through Sequence C, the combined UV/TC/HF test sequence consisting of the UV Preconditioning Test (MQT 10), 50 Thermal Cycles (MQT 11) and 10 Humidity Freeze Cycles (MQT 12). One of those two modules then goes through the Robustness of Termination Tests (MQT 14.1 and MQT 14.2).
- Two modules go through Sequence D consisting of 200 Thermal Cycles (MQT 11).
- Two modules go through Sequence E starting with the Damp Heat Test (MQT 13). Then one of the two modules goes through a static Mechanical Load Test (MQT 16) while the other goes through the Hail Test (MQT 17).

At the conclusion of the different sequences, the stressed modules go through a Final Stabilization Procedure (MQT 19.2) and then the final Post-Test Measurements which consist of Performance at STC (MQT 06.2), the Insulation Test (MQT 03) and the Wet Leakage Current Test (MQT 15).

The pass criteria for IEC 61215: 2016 consists of the following requirements:

- After initial stabilization, the modules must meet the power rating of the name plate – Called Gate 1: Verification of rated label values.
- After each of the stress tests, the power loss shall be less than 5% – Called Gate 2: Maximum power degradation during type approval testing.
- There can be no evidence of open circuits during any of the tests.
- There can be no evidence of "major visual defects" as defined in the document.
- The insulation test requirements shall be met during both pretest and post-test measurements.
- The wet leakage current test requirements shall be met at the beginning and the end of each test.
- Any specific requirements of each individual test shall be met.

The following subsections provide a general description of the tests specified in IEC 61215: 2016.

4.3.1 MQT 01 – Visual Inspection

Visual inspection is performed before and after every accelerated stress test. The visual inspection after one test can be used as the visual inspection before the next test in the sequence. The visual inspection directions say to carefully inspect each module under an illumination of not less than 1000 lux, looking for what are defined as "Major Visual Defects." If any of the items on the list of "Major Visual Defects" is observed, the module is considered to have failed the test. Major visual defects are conditions that could impair performance or safety during the life of the product. Items that are considered Major Visual Defects include things like broken, cracked, torn, or bent surfaces of the module, continuous delamination paths, burnt spots, and removal of a large area of a particular cell due to cracking or corrosion. When performing the IEC 61215 test sequence, it is important to know what is on the detailed list of "Major Visual Defects" found in the document.

4.3.2 MQT 02 – Maximum Power Determination

Maximum power measurements are taken after stabilization and before and after each of the accelerated stress test sequences. They may also be taken before and after each of the accelerated stress tests if desired. Because the change in power is one of the pass/fail criteria, repeatability of the measurement is very important. Equipment required for determining maximum power of the modules includes:

- A solar simulator Class BBA or better in accordance with IEC 60904-9 [14].
- At least one PV reference device meeting the requirements of IEC 60904-2 [15].
- The equipment necessary to take an I-V curve of a module in accordance with IEC 60904-1 [16].

The present version of IEC 61215-2 allows for the I-V curves to be measured at temperatures between 25 and 50 °C and at irradiances between 700 and 1100 W/m^2. The temperatures and irradiances of the measurements before and after stress testing can then be compared using one of the correction methodologies given in IEC 60891 [17].

4.3.3 MQT 03 – Insulation Test

The insulation test is performed as part of the pretest measurements and again as part of the post-test measurements. It is also performed immediately after the Hot Spot Test and after the Robustness of Termination Test.

The insulation test has two parts. In the first part, a high voltage equal to 1000 V plus twice the maximum system voltage of the module is applied between any exposed metal parts of the module and the cell circuit. If there are no exposed metal parts, a foil is wrapped around the outside of the module and one terminal is attached to the foil and the other to the cell string. This voltage is applied for one minute and then the polarity is reversed for one minute. The module passes this part of the test if there is no dielectric breakdown or surface tracking during the application of the high voltage.

In the second part of the test, a voltage of 500 V or the maximum system voltage for the module, whichever is greater, is applied to the module (between exposed metal parts and the cell circuit) for two minutes. Then the insulation resistance is measured. The module passes this part of the test if the measured insulation resistance times the area of the module is equal to or greater than $40\,M\Omega\,m^2$. If the area of the module is less than $0.1\,m^2$ the module passes if the insulation resistance is equal to or greater than $400\,M\Omega$.

4.3.4   MQT 04 – Measurement of Temperature Coefficients

This is one of the characterization tests, not an accelerated stress test. It is performed as the last characterization test in Sequence A and carries no pass/fail criteria. Laboratories are required to report on the temperature coefficients of short-circuit current (α), open-circuit voltage (β) and peak power (δ). Methods for measuring the temperature coefficients are found in IEC 60891. If IEC 61853-1 [18] has been performed on the module type under test, then measurement of the temperature coefficients may be omitted.

4.3.5   MQT 05 – Measurement of NMOT

Measurement of NMOT is another characterization test, not an accelerated stress test and it carries no pass/fail criteria. The NMOT measurement is the first test performed in sequence B. The first two editions of IEC 61215 called for measurement of the NOCT. This was changed to NMOT in the 2016 version. The only difference between the two is that NOCT is measured with the module open circuited, while NMOT is measured with the module biased at or near peak power.

NMOT is defined as the equilibrium solar cell temperature within a module mounted in an open rack under the following environmental conditions:

- Tilt angle: $(37 \pm 5)°$
- Total irradiance: $800\,W/m^2$
- Ambient temperature: $20\,°C$
- Wind speed: $1\,m/s$
- Electrical load: Operated near maximum power point

NMOT is usually measured outdoors using the method given in IEC 61853-2 [19]. If IEC 61853-2 has already been performed on the module type under test, then measurement of NMOT may be omitted.

The NMOT measurement is based on the fact that the temperature difference between a module and the ambient air is independent of the temperature of the air, depending linearly on the amount of solar irradiance incident on the module and on wind that cools the module. The difference between the module temperature and the ambient temperature is measured as a function of solar irradiance and wind speed. The temperature difference data is then fit to Eq. (4.1).

$$T_M - T_{amb} = G / (u_0 - u_1 v) \tag{4.1}$$

Where:

T_M is the module temperature

T_{amb} is the ambient temperature in the vicinity of the module

G is the total solar irradiance incident on the front surface of the module

v is the wind speed

u_0 is a coefficient derived from the data that describes the influence of the irradiance on the temperature difference

u_1 is a coefficient derived from the data that describes the influence of the wind speed on the temperature difference

Since this measurement requires multiple days of data, it is often convenient to perform the NMOT test in conjunction with the Outdoor Exposure Test (MQT 08).

The reason for measuring a module-operating temperature is to provide data to system designers so that they can predict module performance under different operating conditions. Some estimate of module temperature is required to make this calculation. Back in the 1970s, in the early days of PV, the NOCT temperature measured during the qualification test was often the only thermal data on a module type that a system designer had. Today. most system designers have a thermal component of their performance module so there is less reliance on this measurement made during the qualification test.

4.3.6   MQT 06 – Performance at STC and NMOT

Performance at STC and NMOT involve the implementation of the measurement procedure in MQT 02 (Maximum Power Determination) at specified conditions. STC are defined as an irradiance of $1000 \, W/m^2$, a test temperature of 25 °C and a test spectrum of AM1.5 as defined in IEC 60904-3 [20]. STC measurements are required during the pretests immediately after initial stabilization to validate that the modules meet Gate 1 requirements and after final stabilization to validate that the modules meet Gate 2. Once again, the measurements need not be made under exact STC conditions, but must be corrected to STC using IEC 60891 for temperature and irradiance and IEC 60904-7 [21] for spectrum.

Performance at NMOT is designed to provide an estimate of a modules' performance under conditions considered typical for outdoor operations, namely an irradiance of $800 \, W/m^2$, a module temperature of NMOT as measured in MQT 05 and a test spectrum of AM1.5. Performance at NMOT is the first measurement in Sequence A. If IEC 61853-2 has already been performed on the module type under test, then measurement of performance at NMOT may be omitted.

4.3.7   MQT 07 – Performance at Low Irradiance

Performance at Low Irradiance involves the implementation of the measurement procedure in MQT 02 (Maximum Power Determination) at a specified low irradiance condition, namely $200 \, W/m^2$ with a test temperature of 25 °C and the AM1.5 spectrum. It is performed in Sequence A after measurement of performance at NMOT. Measurement of performance at low irradiance is to provide the user with at least one data point on how a particular module type behaves at low irradiance. Today, most PV users will have a model of module performance and will not worry about such a single-point measurement taken on one module. If IEC 61853-1 has been performed on the module type under test, then measurement of performance at low irradiance may be omitted.

4.3.8 MQT 08 – Outdoor Exposure Test

The Outdoor Exposure Test was included in the early days of PV just to validate that a new module type could survive for a minimal length of time outdoors without degradation or deterioration. It is the first test in Sequence B and is shown being performed in conjunction with NMOT (MQT 05). In this test, a module is mounted as it will be deployed outdoors, usually in an open rack and exposed to a cumulative dosage of $60\,kWh/m^2$ as measured by an appropriate irradiance monitor. After the exposure, the module must not have any major visual defects, its wet leakage current must meet the same requirements as for the initial measurements and it can't lose more than 5% of its peak power.

In a sunny location, it usually takes less than three months to reach such a dose so it is unlikely that any reasonably reliable module will fail this test.

4.3.9 MQT 09 – Hot Spot Endurance Test

As indicated in Section 3.7, the Reverse Bias Hot Spot Test is used to determine whether shadows occurring during normal operation will damage the module. The Hot Spot Test is performed at the end of Sequence B (only the bypass diode functionality test is performed after it). The major technical issue is how to identify which cells within the module are most likely to overheat if the module is shadowed. If there are no bypass diodes or the bypass diodes are removable, cells with localized shunts can be identified by reverse biasing the cell string and using an IR camera to observe hot spots. If the module circuit is accessible, the current flow through the shadowed cell can be monitored directly. However, most PV modules have bypass diodes that are not removable and internal electric circuits that are not accessible. Therefore, a non-intrusive method to find the cells most likely to overheat is needed for most module types.

The Hot Spot Endurance Test in IEC 61215-2: 2016 uses an I-V screening method for determining which cells have the lowest and highest shunt resistances within the module. Usually, the worst overheating occurs in cells with either the lowest or highest shunt resistances. This screening method was proposed by Wohlgemuth and Herrmann [22] as early as 2005 but was not adopted into IEC 61215 until 2016. The approach is based on taking a set of IV curves for a module with each cell shadowed in turn. Figure 4.4 shows the resultant set of IV curves for a sample cry-Si module. The curve with the highest leakage current at the point where the diode turns on was taken when the cell with the lowest shunt resistance was shadowed. The curve with the lowest leakage current at the point where the diode turns on was taken when the cell with the highest shunt resistance was shadowed.

For wafer-based modules, the standard calls for selecting the following four cells for testing:

1) The cell adjacent to the edge that has the lowest shunt resistance.
2) The two cells with the lowest shunt resistance in addition to the edge cell selected above.
3) The cell with the highest shunt resistance.

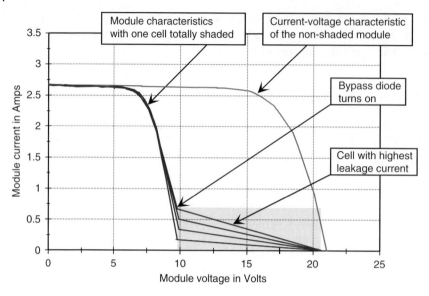

Figure 4.4 Module I-V characteristics with different cells totally shadowed [22].
Source: reproduced from IEEE PVSC with permission from Werner Herrmann of TÜV Rheinland.

The next step is to determine the worst-case shadowing for each of these four cells. This can be done by either of the following methods:

- If the cell circuit is accessible, the current through the shadowed cell can be measured directly. Gradually decrease the shadowed area of the selected cell until the I_{SC} (short circuit current) of the selected cell coincides as closely as possible with I_{MP} (the current through the selected cell when the unshadowed module is producing maximum power).
- If the cell circuit is not accessible, direct measurement of the module short circuit current will not work since the bypass diode will conduct the current around the shadowed cell. The proposed approach is similar to the method utilized to determine the cell shunt characteristics. Take a set of I-V curves which each of the four selected cells shadowed at different levels (for example 100, 75, 50, 25, 10%) as shown in Figure 4.5. From this data determine the worst-case shadowing condition, which occurs when the maximum power point current of the shadowed module coincides as closely as possible with I_{MP} (the current through the unshadowed module at its maximum power).

In both cases, the worst-case shadowing occurs when the bypass diode turns on at a current equal to the unshadowed modules maximum power current I_{MP}. The hot spot test is then conducted at this shadowing level.

Each of the four selected cells is then tested in turn. Short circuit the module, shade each of the four selected cells 100% and use an IR camera to identify the hottest spot on the cell. Then shadow the cell to the worst case as shown in Figure 4.4, trying not to shadow the area with the hottest spot. In this configuration, expose the module to one sun ($1000\,\text{W/m}^2$) at a module temperature of $50\pm10\,°C$ for a minimum of one hour. If the IR camera indicates that the cell temperature is still increasing after the one- hour exposure, continue exposing the cell for a total time of five hours.

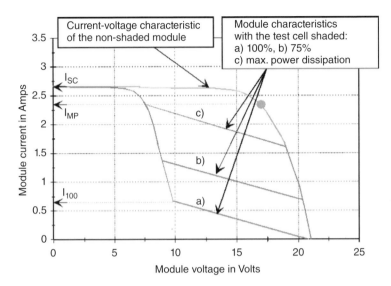

Figure 4.5 Module I-V characteristics with the one of the four selected cells shadowed at different levels [22]. *Source:* reproduced from IEEE PVSC with permission from Werner Herrmann of TÜV Rheinland.

For MLI thin-film modules the Hot Spot Test requires a somewhat different approach. For some MLI modules any shadowing can cause irreversible damage [23]. Because of the geometry of long narrow cells, it is also more likely that multiple cells will be shadowed. Such thin-film modules tend to be produced with different electrical circuitry rather than just with just one series circuit. This complicates the testing in terms of what the preliminary shadowing results look like. The following discussion refers to MLI modules with one series string of cells.

For these modules, IEC 61215-2: 2016 calls for shading an increasing number of cells in series. Maximum power is then dissipated when the reverse voltage across the shaded cells is equal to the voltage generated by the remaining illuminated cells in the worst-case shading condition. This is also when the short-circuit current of the shaded module equals the maximum power current of the non-shaded module. This worst-case shading condition is then applied with an irradiance of $1000\,\mathrm{W/m^2}$ at a module temperature of $50\pm10\,°C$ for one hour.

For other MLI circuit arrangements the concept is similar, that is to find the worst-case shadowing and test with it. The details depend on the current paths through the module. Readers should review the instructions in IEC 61215-2; 2016 and the particular technology specific part (CdTe, a-Si, or CIGS) for guidance.

In both cases (wafer and MLI) the pass/fail criteria require no major visual defects, that the insulation resistance and wet leakage current meet the same requirements as the initial measurements and that the module continues to function. However, there is no requirement for retention of a specific amount of maximum power after the Hot Spot Test. Failure usually occurs when a cell gets so hot that it burns a hole through the encapsulant and backsheet.

4.3.10 MQT 10 – UV Preconditioning Test

It can be very expensive to subject PV modules to long-term UV exposure since they require large area UV sources with long exposure times. Therefore, TC82 (the IEC Technical Committee for PV) has taken the route of subjecting PV modules to only a short UV "pre-conditioning" test before the thermal cycle and humidity-freeze tests at the beginning of Sequence C. This short exposure is designed to identify those materials or adhesive bonds that are highly susceptible to UV degradation. Longer-term UV exposures will then be applied as materials tests where smaller samples can be tested for long times more economically.

The UV preconditioning test conditions are defined as:

UV Light Source – Not specified as long as the other requirements can be met.

Spectral Distribution – No appreciable irradiance at wavelengths below 280 nm, with sufficient irradiance at wavelengths between 280 and 400 nm to provide the necessary dose with at least 3%, but not more than 10% in the wavelength band between 280 and 320 nm. The standard has no requirements for irradiance above 400 nm, so one can use a xenon source with a majority of its energy in the visible or fluorescent lamps with almost no visible light.

Intensity – At wavelengths between 280 and 400 nm, the intensity shall be less than $250 \, \text{W/m}^2$ (i.e. about five times the natural sunlight level), with a uniformity of $\pm 15\%$ over the test plane.

Duration or Dose – Subject the module(s) front side to a total UV irradiation of at least $15 \, \text{kWh/m}^2$ in the wavelength range between 280 and 400 nm.

Type of Cycle – There is no cycle, the UV irradiation is constant.

Temperature of Sample – $(60 \pm 5) \, °C$.

Humidity of Chamber – Not specified.

Sample loading – Module should be biased to approximately peak power or short circuited.

Pass criteria for the UV Preconditioning test include no major visual defects and the wet leakage current meets the same requirements as the initial measurements. After the whole sequence of tests (Sequence C) the modules must pass Gate 2 for maximum power (less than 5% change).

4.3.11 MQT 11 – Thermal Cycling Test

Thermal cycling is performed twice during qualification testing in IEC 61215. Fifty thermal cycles are performed after the UV preconditioning (MQT 10) in Sequence C. Sequence D requires 200 thermal cycles. The thermal cycles used in both sequences are identical.

In the Thermal Cycle test, the modules are placed within a climate chamber that is cycled through a defined thermal profile. At least one module of each type in the chamber is instrumented with an appropriate temperature sensor. During the defined cycle, each module must reach a minimum temperature of $(-40 \pm 2) \, °C$ and a maximum temperature of $(85 \pm 2) \, °C$. The modules must have a minimum dwell time of 10 minutes at both extremes. In both the heating and cooling phases, there is a maximum allowable change in temperature of $100 \, °C/\text{hour}$. Finally, the standard specifies a maximum cycle time of six hours. If a test laboratory wants to use the minimum dwell times and the maximum change in temperature, a cycle time of about three hours is possible. The thermal cycle test can be run anywhere from 4 to 8 cycles per day.

A specified current flow through the module is required during the heat-up cycle. For crystalline-Si modules this current flow is equal to the STC maximum power current for the particular module type. For other technologies, the amount of current flow is defined in the technology specific sections. Typically, the current flow starts as the module is being heated from −40 °C, though this can be delayed until −20 °C if the current flow at −40 °C causes the temperature to rise too quickly (greater than 100 °C/h). The current flow is shut off when the module reaches 80 °C.

To pass the Thermal Cycle test, there can be no interruption of current during the test, no evidence of major visual defects and the wet leakage current must meet the same requirements as the initial measurements. After the whole sequence of tests (Sequence C or D) the modules must pass Gate 2 for maximum power (less than 5% change).

4.3.12   MQT 12 – Humidity-Freeze Test

The Humidity-Freeze test is performed after the 50 thermal cycles in Sequence C. In this test, the modules are placed within a climate chamber capable of applying high humidity while being cycled through a defined thermal profile. At least one module of each type in the chamber is instrumented with an appropriate temperature sensor and a small current is passed through each module to check for continuity. The modules are then placed in the test chamber at room temperature and ambient humidity. They are then heated at a maximum rate of 100 °C/h to (85 ± 2) °C while the humidity in the chamber is increased to $85 \pm 5\%$ RH. The modules are then held at (85 ± 2) °C with a humidity level of $85 \pm 5\%$ RH for a minimum dwell time of 20 hours. The modules are then cooled down with a maximum allowable change in temperature of 100 °C/hour without humidity control until they reach (0 ± 2) °C. Below 0 °C, they can be cooled up to 200 °C/h until they reach (-40 ± 2) °C. They are then held at (-40 ± 2) °C for a minimum of 30 minutes. Finally, they are heated back to room temperature for the completion of one cycle. The time in the cycle between the (85 ± 2) °C soaks should be no longer than four hours.

Sequence C specifies 10 of these humidity freeze cycles. To allow the modules to equilibrate to the environment but not to dry out, a recovery time of two to four hours is specified before subsequent tests are performed.

To pass the Humidity-Freeze test, there can be no interruption of current during the test, no evidence of major visual defects and the wet leakage current must meet the same requirements as the initial measurements. After the whole sequence of tests (Sequence C or D) the modules must pass Gate 2 for maximum power (less than 5% change).

4.3.13   MQT 13 – Damp-Heat Test

The Damp-Heat Test is the first test in Sequence E. In this test, the modules are placed in a climate chamber capable of applying high humidity at elevated temperature. The chamber is then heated to (85 ± 2) °C and humidified to a level of $85 \pm 5\%$ RH. Modules are maintained in the chamber under these conditions for between 1000 and 1040 hours.

To allow the modules to equilibrate to the environment but not to dry out, a recovery time of two to four hours is required before subsequent tests are performed. To pass the Damp-Heat test, there can be no evidence of major visual defects and the wet leakage current must meet the same requirements as the initial measurements. After the whole sequence of tests (Sequence E) the modules must pass Gate 2 for maximum power (less than 5% change).

4.3.14 MQT 14 – Robustness of Termination

The Robustness of Termination Test is performed on one module in Sequence C after completion of the UV Preconditioning (MQT 10), the 50 Thermal Cycles (MQT 11) and the 10 Humidity-Freeze Cycles (MQT 12). It has two parts, a test of the junction box adhesion and a test of the attachment strength of the wires. The initial Junction Box Retention Test is to be performed between two and four hours after completion of the humidity-freeze test.

Measurement of the retention of the junction box on its mounting surface (MQT 14.1) involves applying a force of 40 N directly to the junction, normal to the surface of the module. The force is applied gradually at the center of the junction box for (10 ± 1) seconds in four directions parallel to the edges of the module. To pass the Retention of Junction Box Test, there can be no displacement of the junction box, no evidence of major visual defects and the wet leakage current must meet the same requirements as the initial measurements. After the whole sequence of tests (Sequence C) the modules must pass Gate 2 for maximum power (less than 5% change).

The second part of this test, the Test of Cord Anchorage (MQT 14.2) can be omitted if the junction box is already qualified in accordance with IEC 62790 [24]. The Test of Cord Anchorage (MQT 14.2) involves both pull tests and torque tests of the wires. They are performed after the Retention of Junction Box Test (MQT 14.1) on the same module. If the module comes with wires specified by the manufacturer, they are tested that way. If the junction box is intended to be used with generic cables, a test device has to be built based on the minimum-sized cable specified by the manufacturer. The standard gives some but not much guidance on how to do this.

The unloaded cable is marked so that any displacement relative to the gland can be easily detected. The cable is then pulled for a duration of one second, 50 times, without jerks in the direction of the axis with a force defined in the standard. (The standard gives a diagram of a piece of equipment that can be used to perform this part of the test, but this is only provided as a recommendation). The force varies as a function of the wire diameter going from 30 N for a wire equal to or less than 4 mm in diameter up to a force of 115 N for a cable diameter in excess of 55 mm. To pass this part of the test, the displacement of the cable at the outlet of the junction box cannot exceed 2 mm.

In the next part, the unloaded cable is marked so that any torsion relative to the gland can be easily detected. The cable is then subjected to a torque specified in the standard for one minute. (The standard gives a diagram of another piece of equipment that can be used to apply the torque, but this is only provided as a recommendation). The torque varies as a function of the wire diameter going from 0.1 N for a wire equal to or less than 4 mm in diameter up to a torque of 1.2 N for a cable diameter in excess of 55 mm. To pass this part of the test, the twist or torsion inside the cable gland or other cord anchorage cannot exceed 45°. The cable must be held in position during the test by the cord anchorage.

In addition to the specific requirements in each part of the Test of Cord Anchorage, the pass/fail criteria require no major visual defects and that the insulation resistance and wet leakage current meet the same requirements as the initial measurements. After the whole sequence of tests (Sequence C) the modules must pass Gate 2 for maximum power (less than 5% change).

4.3.15 MQT 15 – Wet Leakage Current Test

The Wet Leakage Current Test is performed before and after every accelerated stress test. The test is performed in a trough or tank of water large enough for the module to be laid flat in the water. The water is treated with a water/wetting agent solution sufficient to wet the surfaces of the module under test and to achieve a resistivity of $3500\,\Omega/cm$ or less. The water temperature must be maintained at $(22\pm2)\,°C$ as the measured leakage current has a strong dependence on temperature.

The cell circuit in the test module is connected to the positive terminal of the voltage supply. The module is then immersed in the water to a depth sufficient to cover all surfaces except junction boxes not designed for immersion (for example, weep holes). If the whole box is not immersed, the part out of the water, as well as any mating connectors should be thoroughly sprayed with the same water solution.

The negative terminal of the voltage supply is then connected to the water in the tank. A voltage of 500 V or the maximum systems voltage of the module, whichever is greater, is then applied between the module circuitry and the water. This voltage level is then maintained for two minutes. Then the insulation resistance is measured.

The module passes the Wet Leakage Current test if the measured insulation resistance times the area of the module is equal to or greater than $40\,M\Omega\,m^2$. For modules smaller than $0.1\,m^2$ the insulation resistance must be equal to or greater than $400\,M\Omega$.

4.3.16 MQT 16 – Static Mechanical Load Test

The Static Mechanical Load Test is performed on one module immediately following the Damp Heat Test in Sequence E. This Mechanical Load Test verifies the minimum test loads that are based on the manufacture's design loads with a safety factor of at least 1.5. The positive and negative design loads may be different (for example, snow load). The design load(s) and safety factors are to be specified in the manufacturer's documentation for each of its approved mounting methods. The minimum required design load per this standard is 1600 Pa which results in a minimum test load of 2400 Pa.

The test module is instrumented to continuously monitor the electrical continuity of its internal circuit during the test. The module is then mounted horizontally on a rigid support structure using one of the mounting methods prescribed by the manufacturer. If there are different possibilities each mounting method needs to be evaluated separately. The module should be held at a test temperature of $(25\pm5)\,°C$. The test load is then carefully and uniformly applied onto the front surface of the module and left there for a minimum of one hour. The test load may be applied pneumatically or by means of weights covering the entire surface. Then the module is turned over and the uplift test load is applied to the backside of the module and left there for a minimum of one hour. This procedure is then repeated three times.

To pass the Static Mechanical Load Test there can be no open circuit detected during the test, no evidence of major visual defects after the test and the wet leakage current after the test must meet the same requirements as the initial measurements. After the whole sequence of tests (Sequence E) the modules must pass Gate 2 for maximum power (less than 5% change).

In the 1993 and 2005 editions of IEC 61215, and the 1996 and 2007 editions of IEC 61646, the static mechanical load level was set at 2400 Pa. To qualify modules for snow loading the last front-loading cycle was increased to 5400 Pa. The present wording still allows this set of parameters if that is what the customer desires. However, modules qualified for snow level via this procedure, continued to fail in the field in regions that encounter heavy snow (See, for example, Figures 2.24 and 2.25). It turns out that such failures are usually not related to a uniform static snow load, but rather to a non-uniform snow load where the snow slides down the front surface of the module and puts a non-uniform load on the frame and/or mounting system. To test for this failure mode, a new IEC standard is under development [25].

4.3.17 MQT 17 – Hail Test

The Hail Test is performed on one module after it has gone through the damp heat test in Sequence E. The first step in the Hail Test is to make ice balls using a mold. The standard provides details on balls from 25 mm up to 75 mm in diameter along with the corresponding mass and test velocity. The minimum size requirement for IEC 61215 is use of 25 mm diameter ice balls. Larger sizes may be used if agreed upon by the laboratory performing the test and whoever is paying for the qualification test. The ice balls are to be made in a freezer at (-10 ± 5) °C and then stored at (-4 ± 2) °C for a minimum of one hour.

The hail test requires the use of a ball launcher that is capable of shooting balls at the module within 5% of the required test velocity, which for a 25 mm ball is 23.0 m/s. The ball velocity should be measured within 1 m of the module surface to an accuracy of ± 2%.

The procedure calls for mounting the modules normal to the direction of ice ball travel in any convenient orientation. The ice balls are to be loaded into the launcher and shot at the module within 60 seconds of being removed from the storage container. A minimum of 11 ice balls are to be shot at the module at the locations given in IEC 61215-2. Shots should hit the module within 1 cm of the stated location.

To pass the Hail Test, there can be no evidence of major visual defects and the wet leakage current must meet the same requirements as the initial measurements. After the whole sequence of tests (Sequence E) the modules must pass Gate 2 for maximum power (less than 5% change).

4.3.18 MQT 18 – Bypass Diode Test

The Bypass Diode Test contains 2 parts. MQT 18.1 is the Bypass Diode Thermal Test that is performed in Sequence B right after the Outdoor Exposure Test. This is an accelerated stress test for the bypass diodes in the module. MQT 18.2 is a Bypass Diode Functionality

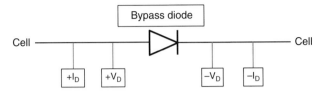

Figure 4.6 Four wire set-up for the bypass diode thermal test.

Test that is used to determine whether the bypass diode is still functioning. The functionality test is performed in Sequence B after the Bypass Diode Thermal Test and again at the end of the sequence after the Hot Spot Test.

The Bypass Diode Thermal Test itself has two parts. In the first part, the temperature of the diode is measured under specified conditions and compared to the manufacturer's rated maximum operating temperature. In the second part, the diodes are exposed to high temperature and current flow to see if they survive. For the temperature measurement part of the test, a special sample can be built to allow for access to the diode. This sample must have the same thermal environment for the diodes as in the module but does not have to contain solar cells. The exposure part of the test must be performed on a real module.

For the measurement part of the test, one of the diodes must be instrumented with a 4-wire system – two wires for applying a current and two wires for measuring a voltage as shown in Figure 4.6. The wires have to be thin enough that they do not contribute significantly to the heat flow out of the diode.

The diode test sample is then placed in a thermal chamber and heated to 30, 50, 70, and 90 °C. While at each temperature, a 1 ms pulsed current equal to the STC short-circuit current of the module is passed through the diode and the voltage (V_D) is recorded. The relationship between V_D and T_J, the temperature of the diode junction is then established using a least-squares fit of the data. Measurement of V_D can be used to determine the diode junction temperature.

The same diode is then heated to (75 ± 5) °C and the STC short-circuit current ($\pm 2\%$) is applied. After one hour, the forward voltage (V_D) of the diode is measured and the V_D versus T_J relationship is used to determine the diode junction temperature under these conditions. To pass this part of the bypass diode thermal test, the measured junction temperature must be less than the diode manufacturer's maximum junction temperature rating for continuous operation.

In the second part of the Bypass Diode Thermal Test, the module is again heated to (75 ± 5) °C. A current flow equal to 1.25 times the STC short-circuit current of the module is passed through the diodes for one hour. To pass this part of the Bypass Diode Thermal Test, there can be no evidence of major visual defects, the wet leakage current must meet the same requirements as the initial measurements and the bypass diodes must still be functional as determined by the Bypass Diode Functionality Test (MST 18.2).

The Bypass Diode Functionality Test can be performed by one of two methods. In Method A, a DC power source's I-V curve tracer is connected to the module at room temperature $(25 \pm 10 °C)$. In this configuration, the current should be passing through the solar cells in the reverse direction and through the bypass diode(s) in the forward direction. The current flow is then increased from 0 to 1.25 times the STC rated short-circuit current of the module and the diode voltage is measured. The diode is still functional if the measured diode voltage V_{FM} meets the relationship in Eq. (4.2).

$$V_{FM} = \left(N * V_{FMrated}\right) \pm 10\% \qquad (4.2)$$

Where:

N is the number of bypass diodes in the module
$V_{FMrated}$ is the diode forward voltage at 25 °C as defined in diode data sheet.

Some modules contain overlapping bypass diode circuits. For such modules, care must be taken to isolate a single diode so that all of the current is flowing through it. If this cannot be accomplished, use Method B instead.

Method B requires taking IV curves with cells shaded as in the Hot Spot Test. In this case, only one cell per diode circuit has to be shaded. You can tell if the bypass diode belonging to the shaded string is working properly, if the characteristic bend in the I-V curve is observed. Many cry-silicon PV modules have three strings with each protected by one diode. In this case, shadowing cells in one string will result in a power drop to roughly two-thirds of the unshaded value.

4.3.19   MQT 19 – Stabilization

IEC 61215: 2016 defines an initial stabilization for the modules that is used to stabilize the modules before the Gate 1 power measurement made before beginning the accelerated stress tests and a final stabilization for Gate 2 measurements made after applying the stresses. In each case, the criterion and procedures are the same. However, both the criterion and the procedures depend on technology and so are given in the technology specific sections of IEC 61215.

The procedure allows for both indoor and outdoor stabilization, except for a-Si which can only be stabilized indoors because of its tendency to anneal at higher temperatures. The indoor procedure requires irradiance levels between 800 and 1000 W/m². Modules are to be maintained at a temperature of (50 ± 10) °C although subsequent restabilizations should be performed within ± 2 °C of the initial stabilization temperature. In the outdoor version, only irradiance levels >500 W/m² count toward the dose requirement. In either case, modules should be biased near their maximum power point during the stabilization process.

For the stabilization process, the initial module peak power (P_1) is measured, the module is exposed to a technology-specific interval of irradiation, the peak power is measured again (P_2), the module is exposed to a second interval of irradiation and the peak power is measured a third time (P_3). The criterion for stabilization is then defined as in Eq. (4.3).

$$\left(P_{max} - P_{min}\right)/P_{average} < x \qquad (4.3)$$

Where:

P_{max} is the largest of the three values P_1, P_2, and P_3
P_{min} is the smallest of the three values P_1, P_2, and P_3
$P_{average}$ is the average of the three values P_1, P_2, and P_3
x is defined in the technology specific sections of the standard

Table 4.4 Technology Specific Stabilization Parameters.

Volume	Technology	x	Initial Stabilization Interval (kWhr/m²)	Final Stabilization Interval (kWhr/m²)
-1	Cry-Si	0.01	5	Not required
-2	CdTe	0.02	20	20
-3	a-Si	0.02	43	43
-4	CIGS	0.02	10	10

Table 4.4 summarizes the technology specific stabilization requirements.

Alternate stabilization techniques can be used for certain technologies. Application of current or voltage bias to solar cells can often lead to effects similar to those of light exposure. Such alternate stabilization procedures should be provided by the module manufacturer along with validation that they yield the same answer as light stabilization.

4.4 How Qualification Tests have been Critical to Improving the Reliability and Durability of PV Modules

The best way to assess how well the Module Qualification Tests have worked is to look at field data. One early study by Whipple [26] looked at module performance during their first 10 years of operation. Module types not qualified to Block V had a 45% failure rate during their first 10 years of operation. For module types qualified to Block V the failure rate in the first 10 years of operation was less than 0.1%. Where would the PV business be today if 45% of the modules failed in the first 10 years of operation? Today, we use IEC 61215: 2016 which is a more comprehensive set of tests than JPL Block V.

Another study was published by Hibberd [27]. This study looked at 125,000 modules from 11 different module manufacturers deployed for up to five years. In this timeframe, only 15 modules failed, a 0.012% failure rate. Over this same time period, 17 modules were damaged (installed next to a golf course) and 89 modules were stolen.

This author also published several articles on field failure rates. Wohlgemuth [28] from the 20th EU PVSEC reported on field returns of Solarex and BP Solar multi-crystalline Si modules deployed from 1994 to 2005. Over the course of the 11 years, the warranty return rate was 0.13% or one failure for every 4200 module years of operation. Wohlgemuth [29] from the 23rd EU PVSEC reported on field returns of Solarex and BP Solar multi-crystalline Si modules from 2005 to 2008 and found an annual return rate of 0.01%.

To understand why the qualification tests have been so important to PV module reliability, it is best to know how the qualification tests were and still are used by module manufacturers. Before there was a qualification test sequence (back in the 1970s) the only way to evaluate the reliability of new module types and designs was to deploy them outdoors for testing. It usually takes a long time for modules to degrade significantly outdoors, especially as they become more reliable. A manufacturer cannot wait years to see how long a new module type is going to survive or whether the new design is going to last longer than the old design before beginning to sell them. Therefore, in parallel with any outdoor testing, modules were being offered for sale. This often led to the types of early field failures reported in the Whipple paper – 45% not surviving for 10 years.

Once the use of Qualification Tests became routine, module manufacturers could use these tests on their new products. Most of the Block Buy module types did not pass the qualification tests on the first try, but once a manufacturer had a design that did, they had a baseline against which all new products could be tested. Once a manufacturer develops a new product, it is put through the qualification test sequence. Since the tests only take a few months they can now wait until the new product passes the qualification tests before beginning to manufacture and ship them. This process eliminates many of the causes of infant mortality so as the paper by Whipple showed, failure rates drop dramatically.

Another interesting measure of a PV module's reliability is the manufacturer's warranty. While setting the warranty is not really a technical activity, but rather a commercial decision, in most cases, a manufacturer will not drastically overpromise on warranty because of the consequences associated with having to replace large volumes of modules. Therefore, warranties did not start to become extended until module manufacturers became comfortable with the use of accelerated stress tests in general and the Qualification Tests in particular to assess module reliability.

Over the years, different companies offered different warranties, but during the 1980s through the early 2000s most of the crystalline silicon module manufacturers offered similar warranty terms. Therefore, the warranty history will be provided for Solarex Corporation (for its cry-Si power modules) with the understanding that other module manufacturer's warranties would be similar but certainly not identical. Module warranties typically have three parts:

1) A workmanship component saying that the module would continue to look nice and not exhibit defects during the specified time period, originally one year, but now typically 10 years.
2) An intermediate power guarantee, initially half the final power level in half the time for the final power guarantee. Today, however, this intermediate power guarantee takes many forms including linear yearly steps over the full warranty period.
3) A final power loss guarantee, typically no more than 20% power loss over the length of the warranty.

Table 4.5 shows the length in years of the Solarex Corporation Final Power Warranty for Crystalline Silicon Power Modules. The year before Solarex increased its warranty from 10 to 20 years, its major competitor; Siemens Solar increased their warranty from 10 to 15 years. Solarex didn't just match the Siemen's warranty, but offered a longer one than Siemens. In this case, the Solarex technical staff was able to provide the technical justification for increasing the warranty to 20 years [30]. As described in this reference, the justification was based on both field data and accelerated testing of modules beyond the qualification levels. The critical issue with a warrant is whether the manufacturer can provide data from the field and from accelerated stress tests that validates the claim that their modules can survive at least as long as the warranty period.

Table 4.5 Solarex warranty period for crystalline silicon modules.

Date	Length of Warranty
Before 1987	5 Years
1987 to 1993	10 Years
1993 to 1999	20 Years
After 1999	25 Years

A final proof of how well the Qualification Tests work is the PV industry itself. The dramatic growth of the industry as outlined in Chapter 1 would not have been possible without at least a perception that PV modules are reliable. We would not be seeing billion-dollar investments in PV modules if they were not many examples of PV modules performing reliably for extended time periods.

4.5 Limitations of the Qualification Tests

While we have seen how useful the Qualification Tests can be, it is important to note that by design, they have limitations. In order to be completed in a reasonable amount of time, the durations of the accelerated stress tests in the Qualification Test Sequence are limited. In addition, in order to have strict pass/fail criteria, there is no attempt to distinguish between a module type that just passes with one that shows no degradation at all. Qualification tests are not designed to:

- Identify and quantify wear-out mechanisms. The tests used are of too short a duration to cause the modules to wear out. Chapter 9 will address what has to be done to predict module service life.
- Address failure mechanisms for all climates and system configurations. The Qualification Tests are based on specific failure modes observed in the field and subsequently duplicated via one or more of the accelerated stress tests within the Qualification Test Sequence. Over time, observations of new failure modes lead to development of new tests that can ultimately be incorporated into the Qualification Test Sequence. Specific examples of the present-day efforts to do this will be discussed in Chapter 7.
- Differentiate between products that may have long and short lifetimes. The Qualification Tests usually help to eliminate infant mortality but say little about which type of module will survive longer. Longer-term accelerated stress tests are required to do this. Chapters 7, 9 and 10 will discuss the development of methods that can be used to differentiate products.
- Address all failure mechanisms in all module designs. Once again, the Qualification Tests only address a specific set of tests known at the time they were developed. New editions attempt to add new tests to cover failure modes that were not evaluated in the older versions. Chapter 7 will talk about the Photovoltaic Quality Assurance Task Force (PVQAT) effort to identify and test for additional failure modes.

- Quantify lifetime for different applications or climates. Because different climates and methods of mounting result in different stress levels on modules, one set of tests is unlikely to ever be able to provide an answer for all situations. This will be discussed further as part of both the PVQAT effort in Chapter 7 and under Service Life Prediction in Chapter 9.

4.6 PV Module Safety Certification

In addition to IEC standards for Module Qualification, there are also Module Safety Standards. It is important that modules be capable of performing safely over the course of their lifetime. This ability to continue safety operation may be different from qualification which is designed to measure the module's ability to continue to produce electricity. The first module safety standard was established by the Underwriters Laboratories (UL), UL 1703, first published in 1986. It addresses electrical, mechanical and fire safety and uses accelerated stress tests similar to those in JPL Block V to evaluate whether the modules remain safe during their years of operation. An important distinction between UL 1703 and Block V or IEC 61215, is that after the accelerated stress tests in UL 1703, the modules must be safe but do not have to perform to specification – they don't have to make any electricity at all to pass the safety tests. Therefore, neither UL 1703 nor any other module safety standard, are good replacements for the qualification tests. The safety certifications tell you nothing about module reliability or durability. This is important to realize because there are certain locations, particularly in the US, where the only standard that modules must meet is UL 1703. So, it is up to the customer to require that the module carries certification to a Qualification Standard, hopefully IEC 61215.

IEC addressed module safety after the publication of IEC 61215. Safety is, in general, more of a national (or regional) issue than qualification or performance, so trying to develop a consensus safety document that meets the needs of most nations around the world can be difficult. For example, in the US, per the National Electric Code (NEC) all electrical systems are grounded. In Europe, most electrical components are double insulated and the systems are not grounded. Trying to write a component safety standard that can meet both needs was very difficult. After many years of trying, IEC finally published a module safety document, IEC 61730 in 2004 and a second edition in 2016 [31].

UL 1703 has been written into the US NEC and so is required by law in most places within the US. The concept for many years has been to try to harmonize UL 1703 with IEC 61730. There were too many differences between the first edition of IEC 61730 and UL 1703, so the harmonization effort had to wait until the second edition of IEC 61730 was published in 2016. Finally, in December 2017, UL adopted a US-version of IEC 61730 to replace UL 1703 (called UL 61730). In the US, PV modules may now be safety certified to UL 61730 though they can still be certified to UL 1703. As of December, 2019, all new module safety certifications in the US must use UL 61730. Since UL 1703 is going away, the remainder of this chapter will focus on describing the requirements in IEC 61730-1 and IEC 61730-2 as well as to point out the US National Differences defined in UL 61730.

Different from Qualification Tests, IEC 61730 (and UL 1703) contain both construction requirements (IEC 61730-1) and test requirements (IEC 61730-2). These will be discussed in the two subsections below. Once again, these descriptions are the author's summary to familiarize readers with what the standard requires and should not be used for performing the tests.

4.6.1 Construction Requirements: IEC 61730-1

IEC 61730-1 specifies and describes the fundamental construction requirements for PV modules necessary to provide safe electrical and mechanical operation. Protection is to be achieved by combinations of constructional measures used to build the module, together with instructions on how the module is to be installed. Many of the requirements for PV modules are taken from other parallel IEC standards and applied to PV modules based on their use and environment. The requirements for module construction start out very general and then drill down to more detailed requirements for the incorporated components and materials.

PV modules should be designed for operation in outdoor non-protected locations, which according to IEC 61730-1, means exposure to solar radiation, in an ambient temperature range of at least −40 to +40 °C and up to 100% relative humidity as well as to rain, hail, snow and sleet. Incorporation of a PV module into the final assembly cannot require any alteration of the PV module from its originally evaluated form. Modules can only be mounted and wired, using methods specified in the installation instructions provided by the module manufacturer. The methods specified in the manufacturer's installation documentation must have been evaluated for compliance with the IEC 61730 series of standards. Any other mounting methods violate the safety certification. Many of the general construction requirements like equipotential bonding continuity, sharpness of edges and potential for loosening of connectors are evaluated for compliance by testing incorporated into part 2 of this standard (IEC 61730-2).

4.6.1.1 Components

Modules typically contain a number of components. Part 1 provides a list of potential components along with requirements for each.

Internal wiring: Internal wiring within a module (for example, cell interconnects) must be sized for the application, sufficiently corrosion-resistant based on the encapsulation package and insulated if it is possible for such an internal conductor to contact parts of more than one polarity. This can happen, for example, when module design calls for internal connectors to pass behind cells. Compliance with these requirements is checked via visual inspection and the Reverse Current Overload test in IEC 61730-2.

External wiring and cables: External wires and cables are required to conform to IEC 62930 [32].

External Connectors: External DC connectors are required to conform to IEC 62852 [33].

Junction boxes for PV modules: Junction boxes for PV modules are required to conform to IEC 62790 [34].

Frontsheets and backsheets: Frontsheets and backsheets are the outer layers of the module and provide protection to keep water and dust out and the electric potential in.

They usually serve as the "relied upon insulation" for the module and, as such, must be capable of withstanding relevant mechanical, electrical, thermal and environmental stresses. Compliance is demonstrated via the tests in IEC 61730-2. While frontsheets and backsheets can be made from a variety of materials including glass, many are polymeric. For polymeric frontsheets and backsheets, there is a Technical Specification, IEC 62788-2 [35] that provides guidance on their preferred properties, as well as a set of requirements for "Polymeric materials used as electrical insulation" from IEC 61730-1 that include the following:

- Endurance to electrical stress: Materials used as electrical insulation should be capable of withstanding the electrical stresses that occur in the application both in the unconditioned state and after exposure to the accelerated stresses used in IEC 61730-2.
- For evaluation of creepage and clearance distances (see subsection later in this chapter), each material is ranked according to its comparative tracking index (CTI) as defined in IEC 60112 [36]. This value rates how easily an electric arc tracks along the surface of the material. Greater distances are required if the tracking is easy.
- Whenever electrical stress is present through a material layer (not along an interface or surface) the concept of distance through insulation is applicable. The international version of IEC 61370-1 has a requirement called "Insulation in Thin Layers", which requires the frontsheet and backsheet to have a minimum thickness based on the rated voltage of the module. The US has eliminated this requirement in UL 61730-1 believing that the high-voltage tests in IEC 61730-2 are much better indicators of the ability of the insulators to withstand high voltage than a requirement for a particular film thickness.
- The materials used in frontsheets and backsheets should have a minimum relative thermal endurance equal to or greater than the maximum normalized operating temperature of the material as determined in the temperature test (MST 21 in IEC 61730-2), or 90 °C, whichever is higher. This value is determined using either the relative thermal index or temperature index (RTE, RTI, or TI) in accordance with IEC 60216-5 [37], UL 746B [38] or IEC 60216-1 [39]. These tests are designed to validate that the material can operate at the rated temperature for a long time period. The RTE value is usually based on 20 000 hours or approximately 2.25 years of testing. After the thermal exposure, the material's mechanical and electrical properties must not have degraded by more than 50% from their initial value.

Insulation barriers: An insulation barrier is a raised or recessed portion of an insulator designed to increase the creepage distance (on the surface) between conducting points. In PV, these are most often seen in the junction box where they increase the creepage distance between terminals. Insulation barriers should be capable of withstanding all of the relevant mechanical, electrical, thermal, and environmental stresses a module is likely to endure. If they are polymeric, insulation barriers should meet the requirements for polymers and they should have a minimum relative thermal endurance equal to the operating temperature of the material as determined in the temperature test (MST 21 in IEC 61730-2) (Similar to the RTI/RTE requirements for backsheets and frontsheets).

Electrical connections: Electrical connections should be designed so that contact pressure is not transmitted through insulating material other than hard materials such as ceramics, unless there is sufficient resilience in the metallic parts to compensate for any shrinkage or movement of the insulating material, for example, by using a spring contact. Loosening of all electrical connections should be prevented, for example, by using a lock washer. Precautions should be taken to ensure that clamping units or other terminations do not loosen due to thermal or mechanical stress during use.

Terminals for external electrical connections should be suitable for the type and range of the conductor's cross-sectional area in accordance to the manufacturer's specification. They should meet the requirements of the junction box standard IEC 62790. Insulated terminals should also be designed to prevent any reduction in clearances and creepage distances.

Connections inside a PV module should be mechanically secured and provide electrical continuity. Electrical connections can be made any way the manufacturer wishes, as long as they are securely connected. A soldered or conductive adhesive joint must also be mechanically secured. Being embedded in the encapsulant is considered an adequate means of mechanically securing of soldered and conductive adhesive electrical connections in a PV module.

Compliance of the electrical connections are checked by the tests of IEC 61730-2 including the visual inspection (MST 01), continuity test of equipotential bonding (MST 13) and screw connection test (MST 33), if it is applicable.

Encapsulants: Encapsulants are considered as a part of the laminate and so are usually not tested separately but as part of the laminate using the tests of IEC 61730-2. However, there is no reason that they could not be tested for and then used as part of the "relied upon insulation." The technical properties of encapsulant should be suitable for the intended application including:

- the rated operating temperature range of the encapsulant should include the temperature range of the module in the intended application, and
- the material group (CTI rating), the insulation resistance and the dielectric strength should all be suitable for the module's intended application.

Compliance of the encapsulant to these requirements is validated by passing the test sequence in IEC 61730-2.

Bypass diodes: Bypass diodes must be rated to withstand the current and voltage for their intended use. Compliance of the bypass diodes to these requirements is validated by passing the following tests from IEC 61730-2:

- bypass diode thermal test (MST 25),
- hot-spot endurance test (MST 22),
- bypass diode functionality test (MST 07), and
- visual inspection (MST 01).

4.6.1.2 Mechanical and Electromechanical Connections

There are numerous mechanical connections within a module, for example, those holding the pieces of the frame together, attaching the junction box and connecting the module to the mounting structure. Mechanical connections should be able to withstand the thermal, mechanical, and environmental stresses that occur in each application and still maintain their functionality. Parts intended to be removed during installation or maintenance, including junction box lids, should require a tool to detach. If the tool is used correctly, it should not come in contact with active live parts within the module. For mechanical connections friction between surfaces, such as simple spring pressure, is not acceptable as the sole means to inhibit turning or loosening of the part.

Compliance of the mechanical connections is checked by the following tests in IEC 61730-2:

- Visual inspection (MST 01)
- Mechanical load test (MST 34)
- Module breakage test (MST 32)
- Materials creep test (MST 37); and
- Continuity of equipotential bonding tests (MST 13).

Some specific mechanical fastening methods have additional requirements that are given below.

Screws: Screws within a PV module should not be made of materials that are soft or are liable to creep. Screws that are operated for maintenance purposes should not be made of insulating material if their replacement by a metal screw could provide a conductive path where one is not intended. To electrically bond the frame pieces together, there should be at least one screw connection (or other electro-mechanical connection means) between each of the metallic components.

Screws with a nominal diameter of less than 3 mm that are used for mechanical and electrical connections should only be used to screw into metal. Screws used for mechanical and electrical connections should engage at least two full threads into the metal. Screwed and other fixed connections between different parts of the PV module should be made in such a way that they do not come loose through torsion, bending stresses, vibration, etc., that may occur in normal use.

Compliance of screw connections is validated by those tests from IEC 61730-2 given for mechanical connections plus the screw connection test (MST 33).

Thread-cutting screws: Thread-cutting screws and self-tapping screws should not be used for the interconnection of current-carrying parts made of soft metal or if the metal is likely to creep. Thread-forming screws (for example, sheet-metal screws) can only be used for the connection of current carrying parts, if they clamp these parts directly in contact with each other, and are provided with a suitable locking means.

Thread-cutting (self-tapping) screws can only be used for the connection of current-carrying parts if they generate a full form standard machine screw thread. These types of screws cannot be used if they are likely to be operated by a user or installer.

Thread-cutting and thread-forming screws can be used to provide continuity for equipotential bonding, as long as it is not necessary to disturb the connection in normal use. For equipotential bonding, one screw is permitted as long as two full threads engaged the metal.

Rivets: Rivets that serve as both electrical and mechanical connection must be locked against loosening, for example, by using a noncircular shank or a notch.

Connections by adhesives: Adhesives are often used to make mechanical bonds in PV modules, for example, to attach the junction box to the backsheet or the laminate to the frame. Compliance for mounting adhesives is checked using the following tests from IEC 61730-2:

- Mechanical load test (MST 34),
- Test of continuity of equipotential bonding (MST 13); and
- Module breakage test (MST 32).

Compliance for junction-box adhesives is checked with the following tests from IEC 61730-2:

- Robustness of termination test (MST 42) and
- Wet Leakage Current Test (MST 17).

4.6.1.3   Materials

Several specific types of materials (polymeric, metallic, and adhesives) have special requirements that are detailed in the following subsections.

4.6.1.3.1   *Polymeric Materials*

Polymeric materials, no matter where or how they are used in the module, should be able to safely withstand the electrical, mechanical, thermal, environmental, and corrosive stresses occurring in the application. Their electrical and mechanical properties should not degrade more than an allowable amount during the lifetime of the module. IEC 61730-1 provides a list of functions that polymeric materials can have in a module and for each of these functions provides a list of requirements. However, the lists are quite confusing and appear to leave out some critical tests in some categories for no apparent reason. For example, why isn't a polymeric material used for an external surface required to be rated to survive the temperatures that modules are likely to see? So, the following list of requirements may not match exactly with IEC 61730-1, but is a logical summary of what properties are most likely to yield a safe module construction.

Polymeric parts which ensure the electrical and/or mechanical safety of the PV module should meet the requirements of the materials creep test (MST 37) from IEC 61730-2. It is hard to image a polymeric material that does not contribute to either or both the electric and mechanical safety of the module so it would be better to just say that all modules should be able to pass the module creep test regardless of what functions the polymeric materials are designed to meet. In addition, remember that the polymeric materials will be stress tested at the module level in IEC 62730-2 by performing, the Insulation Test (MST 16) and the Wet Leakage Current Test (MST 17) before and after all of the accelerated stress tests and the Impulse Voltage Test (MST 14).

Polymeric materials should be capable of surviving the weathering stresses occurring in the application. IEC TC82 has recently published a technical specification for evaluation of the weathering capabilities of polymeric materials, IEC TS 62788-7-2 [40]. The second edition of IEC 61730-1 should clarify how this test is to be used to evaluate the weathering of the polymeric materials within the modules. IEC TS 62788-7-2 will be discussed further in Chapter 7 under the PVQAT effort and in Chapter 9 on predicting module service life.

Polymeric materials used for insulation, external parts or that provide mechanical functions should have a minimum relative thermal endurance (RTE, RTI, or TI) equal to or greater than the maximum normalized operating temperature of the material as determined in the temperature test (MST 21 in IEC 61730-2), or 90 °C, whichever is higher, as was described above, for backsheets and frontsheets. The only difference could be which property is evaluated before and after the thermal exposure. If the material only provides insulation properties then the thermal endurance can be evaluated based on electrical properties. If the material only provides a mechanical function, then the thermal endurance can be evaluated based on a mechanical property. Often, however, like a backsheet or frontsheet both properties are important and so both should be tested before and after the thermal exposure.

As discussed above for backsheets and frontsheets, polymeric materials that provide part or all of the relied upon insulation are part of the evaluation of creepage and clearance distances (see subsection later in this chapter), and so each should be ranked according to its CTI.

Finally, materials that will be exposed to the outside world during use must be rated for flammability using IEC 60695-11-10 [41] and the ball pressure test using IEC 60695-10-2 [42]. The required class for flammability and the temperature of the ball pressure test vary depending on the function of the material within the module. IEC 61730-1 also calls for testing insulation in thin layers using the ignitability test (MST 24) from IEC 61730-2. The US has eliminated this requirement in UL 61730-1, because in the US, all modules intended for installation integral with or forming a part of the building's roof structure must be evaluated in accordance with the Standard Test Methods for Fire Tests of Roof Coverings, UL 790 [43].

4.6.1.3.2 Metallic Materials

Metal parts to be used in places with wet or humid climates should be protected from corrosion. Of course, almost all places that PV modules are deployed experience some rain and even in the desert moisture condenses on modules at night. So, we can interpret the requirements to mean that unless a metallic component is protected from the environment by a sealed package (for example, using a glass/glass laminate with edge seals), the metal itself should be protected by the appropriate coating such as plating, painting, or enameling. The corrosion protection should be at least equivalent to a zinc coating of 0.015 mm in thickness, although the standard does not indicate how this can be measured or evaluated. Simple sheared or cut edges and punched holes are not required to be coated.

In addition to the coating requirements, there is also a restriction on what different metals can be in contact with one another. In accordance with IEC 60950-1 [44] metal parts should not be in contact with other metal parts that have a difference in electrochemical potentials of more than 600 mV. The material combinations listed in Table J.1 of IEC 60950-1: 2005 are there to serve as a guideline to determine electrochemical potentials between two dissimilar materials and therefore which metals can be used in contact with each other.

Current-carrying metallic parts should have sufficient mechanical strength and electrical conductivity to serve the function they are designed for. Once again, unless they are packaged in a sealed environment, they should be considered susceptible to corrosion and therefore should be protected against corrosion, e.g. by coating. This means that most of the interconnect ribbons and bus bars within the laminate are usually tin- or solder-plated to make them corrosion resistant. These coatings should comply with one the following standards – ISO 1456 [45], ISO 1461 [46], ISO 2081 [47] or ISO 2093 [48].

4.6.1.4 Protection Against Electric Shock

Protection against electric shock is one of the most important safety functions of the PV module and of this IEC safety standard. IEC has a long history of developing standards for protection against electric shock so the IEC PV technical committee (TC82) has applied these standards to PV in establishing the requirements within IEC 61730. To help understand these requirements, let's start out by defining several terms.

Insulation is used to protect personnel from high voltages within a piece of equipment. IEC has defined the following three different types of insulation based on their usage.

- Functional insulation: Insulation that is necessary for the proper functioning of the equipment but does not protect against electric shock.
- Basic insulation: Insulation of high-voltage parts which provides basic protection against electric shock.
- Double insulation: Insulation comprising both basic insulation and supplementary insulation that provides extra protection against electric shock.

The original idea behind double insulation was that it consists of two separate layers of insulation so that if one failed the other would still retain the necessary level of safety. Today, this has been modified somewhat as the standards also allow reinforced insulation, which provides a degree of protection against electric shock that is considered equivalent to double insulation. So, the real distinction between basic and double insulation will be found in the construction requirements (that is the required spacings within the product) and the test levels specified for each.

Protection against electric shock will be achieved by combinations of constructional measures used to build the module, together with how the module is installed. PV modules are classified according to IEC 61140 [49]. The Classes as they apply to PV modules are defined as:

Class 0 – These are high-voltage and high-power modules that are used in restricted locations (for example, behind the fence in a utility scale facility). These modules must be provided with basic insulation protection.

Class I – There are actually no such things as Class 1 PV modules. Class 1 requires the equipment to be surrounded by a grounded conductor. Since at least the front of the PV module must be exposed to sunlight, it cannot be covered by such a grounded conductor, at least until we discover a way to make bulk optically-transparent electrical conductors.

Class II – These are high-voltage and high-power modules that are used in generally accessible areas where human contact with the outside of the module is possible. These modules must be provided with double or reinforced insulation.

Class III – These are low-voltage and low-power modules that are considered inherently safe. IEC 61730-1 defines the limits as open-circuit voltage of less than 35 V, STC peak power of less than 240 W and short-circuit current of less than 8 A. In the US, UL 61730 revises the limit for open-circuit voltage to less than 30 V. Because they are considered

Table 4.6 Required type of insulation for different classes of modules from IEC 61730-1.

Module Protection Class	Touch Protection Required	Insulation between live parts and accessible metal parts	Insulation between live parts and outside world	Insulation between different live parts within the module
Class 0	Yes	Basic	Basic	Basic
Class II	Yes	Double	Double	Basic
Class III	No	Functional	Functional	Functional

inherently safe, Class III modules can be used anywhere as long as the limits of voltage, power and current are not exceeded. They do not require basic or reinforced insulation although Class III modules may require functional insulation.

Table 4.6 summarizes the types of insulation required for the different functions within modules of a specific class. This table defines the type of insulation which establishes construction requirements in IEC 61730-1 and the test-voltage levels in IEC 61730-2. Neighboring solar cells connected in series have no special insulation requirements as long as the maximum power dissipation between the two cells is less than 15 W (based on the rating of the solar cells).

The construction requirements for the insulation system in a PV module depend on a number of factors, including the following:

- Overvoltage category from IEC 60664-1 [50], which is based on the transient voltages that a PV module may see during its use in a power system. Since PV modules are used in many different types of systems and applications, the project team for IEC 61730 determined PV modules should be Overvoltage Category III. This decision impacts the selection of the impulse voltage test level (MST 14) and the required clearance distances within the module. If modules were only used in more protected systems, the Overvoltage Category could be lower, reducing the impulse voltage test levels and shrinking the required clearance distances.
- Pollution degree (PD), which is a measure of the level of the expected moisture and dust present in the micro-environment. In this case, the micro-environment means inside the different parts of the PV module. One of the approaches to modifying the construction requirements is to keep pollutants out and improve the PD. The outdoor environment that PV modules are used in is considered PD 3. For enclosures having a degree of protection IP 55 or higher according to IEC 60529 [51] the PD can be reduced to 2. This can be validated for PV modules by successfully passing all of the accelerated stress tests in IEC 61730-2 except for Test Sequence B-1. Of course, to be safety certified a PV module must pass all of these tests so, in reality, any PV module certified to IEC 61730 can be considered PD 2 or better. If the module type also passes Test Sequence B-1, then the module type can be considered PD 1.
- Properties of the materials used in the construction. These materials must be able to withstand the voltages expected during operation. This is validated using the high-voltage tests of IEC 61730-2. If the surfaces of a construction material allow a tracking path between points of different voltages (either within or outside the module) their CTI must

be measured using the procedure given in IEC 60112. The better the CTI Material Group rating for a material is (I is the best and III is the worst) the closer together the voltage sources can be along its surface.

- Rated systems voltage, which is the highest voltage that the manufacturer rates its module for use at. Today, many module types are rated for 1000 V. IEC 61730 allows for rating of modules up to 1500 V.

In IEC 61730-1, these factors are all combined to provide the required values for creepage, clearance and distance through solid insulation. The creepage distance is the shortest distance along the surface of a solid insulating material between two live parts or between a live part and accessible or exposed parts. The clearance distance is the shortest distance through air between two conductive parts, or between a conductive part and an accessible surface. In IEC 61730-1 these values are given in two large tables, one for basic insulation, applicable to Class 0 and III, and one for double or reinforced insulation, applicable to Class II. Rather than try to reproduce these complicated tables in this book, it will discuss some of the typical applications of the tables. The readers should always consult IEC 61730-1 before trying to qualify a module type to the module safety standard.

Let's start by looking at Class III requirements for the inherently safe modules (>35 V) for low-power applications. These are used in single module off-grid systems.

<u>Class III – Minimum Thickness of Thin Layers (DTI) = 0.01 mm</u> (But remember this is not required in the US). So, internationally, this means the frontsheet and backsheet of a Class III module must be at least 0.01 mm thick.

<u>Class III – Required distance between internal live parts and any outer accessible surface and between internal live parts at different potentials inside PV module</u>:

The Minimum Clearance distance = 0.1 mm for PD 1, 0.2 mm for PD 2 and 0.8 mm for PD 3.

The Minimum Creepage distance is given in Table 4.7a.

<u>Class III – Required distance between terminals of different potentials inside a rewireable junction box</u>:

The Minimum Clearance distance = 0.5 mm.

The Minimum Creepage distance is given in Table 4.7b.

The values for "Allowable distance between internal live parts and accessible surfaces" define how close a part of the active circuit (cells, bus bars, etc.) can come to the edge of the laminate. We see from these construction requirements that using PD 2 values (remember

Table 4.7a Minimum Creepage values for Class III modules.

Pollution Degree (PD)	Materials Group		
	I	II	III
1	0.2 mm		
2	0.6 mm	1.0 mm	1.2 mm
3	1.5 mm	1.7 mm	1.9 mm

Allowable distance between internal live parts and accessible surfaces and between internal live parts at different potentials inside PV module.

Table 4.7b Minimum Creepage values for Class III modules.

Pollution Degree (PD)	Materials Group		
	I	II	III
1	0.4 mm		
2	1.2 mm	1.7 mm	2.4 mm
3	3.0 mm	3.4 mm	3.8 mm

Allowable distance between terminals of different potentials inside a rewireable junction box.

any module passing the IEC 61730-2 tests except the B-1 sequence are no worse than PD 2), you have a value of 0.2 mm for clearance and no more than 1.2 mm for creepage for Materials Group III reducing to 0.6 mm for Materials Group I.

Now let's look at a category of products that probably covers the vast majority of those sold in the world, namely Class II modules with double insulation rated at 1000 V. These can be used in most utility scale plants and on buildings.

Class II 1000 V – Minimum Thickness of Thin Layers (DTI) = 0.15 mm (But remember this is not required in the US). Internationally, this means that the frontsheet and backsheet of a Class II 1000 V module must be at least 0.15 mm thick. This may be thicker than is conveniently available for polymeric frontsheets.

Class II 1000 V – Required minimum distance between internal live parts and any outer accessible surface and between terminals of different potentials inside a rewireable junction box:

The Minimum Clearance distance = 14 mm
The Minimum Creepage distance is given in Table 4.8a.

Class II 1000 V – Minimum required distance between internal live parts of different potentials inside the PV module:

The Minimum Clearance distance = 8 mm
The Minimum Creepage distance is given in Table 4.8b.

Table 4.8a Minimum Creepage values for Class II modules rated for 1000 V.

Pollution Degree (PD)	Materials Group		
	I	II	III
1	6.4 mm		
2	10 mm	14.2 mm	20 mm
3	25 mm	28 mm	32 mm

Allowable distance between internal live parts and accessible surfaces and between terminals of different potentials inside a rewireable junction box.

Table 4.8b Minimum Creepage Values for Class II Modules rated for 1000 V.

Pollution Degree (PD)	Materials Group		
	I	II	III
1	3.2 mm		
2	5.0 mm	7.1 mm	10 mm
3	12.5 mm	14 mm	16 mm

Allowable distance between internal live parts of different potentials inside the PV module.

Reviewing these construction requirements, we see that it is really important to be able to use PD 1 or 2 values or the edge isolation requirements become excessive. Also note that outside the US, the distance through insulation requirement starts to impose restrictions as there are many constructions with backsheets and frontsheets thinner than 150 µm that have been successful in passing the 61730-2 tests and surviving in the field in high-voltage systems for many years. This demonstrates why it is so important to develop a real dc breakdown voltage test for backsheets and frontsheets rather than relying on thickness.

According to IEC 61730-1 one more way to reduce the required edge isolation distance is by using a cemented joint. A cemented joint is defined as a joint between two insulating materials where the interface has been tested to demonstrate that it is cemented, and thus can be considered as solid insulation with no interface for creepage. To test for a cemented joint, you complete the tests of IEC 61730-2 with the following changes:

1) In the Visual Inspection (MST 01) there can be no observations of cracks or voids within the cemented joint.
2) In the Insulation Test (MST 16) the test voltage is increased to 1.35 times the standard.
3) In the Wet Leakage Current Test (MST 17) the test voltage is increased to 1.35 times the standard.
4) The insulating adhesive/sealant should have a volume resistivity of greater than $50 \times 10^6 \, \Omega$-cm (dry) and greater than $10 \times 10^6 \, \Omega$-cm (wet) using IEC 62788-1-2, test method A [52].
5) For rigid/flexible and all flexible joints, the construction must pass the Peel Test (MST 35).
6) For rigid/rigid joints, the construction must pass the Lap Sheer Test (MST 36).

If a module is determined to have cemented joints then the required minimum distance through the cemented joint is given by the values in Table 4.9. These are small values that allow the active electric circuit to be extended quite close to the edge of the module.

Qualification of a module's construction to IEC 61730-1 is the first step in getting a module type safety certified. The module manufacturer must know what Class and what system voltage rating they are designing the product for. An accredited test laboratory will then review the module construction versus the requirements of Part 1. If all of that is okay, they go on to the tests given in Part 2.

Table 4.9 Minimum required distance through cemented joints.

Rated Voltage	For Class 0 & III	For Class II
<35 V	0.1 mm	0.2 mm
100 V	0.2 mm	0.3 mm
150 V	0.25 mm	0.5 mm
300 V	0.5 mm	1.0 mm
600 V	0.7 mm	1.5 mm
1000 V	1.0 mm	2.0 mm
1500 V	1.7 mm	3.5 mm

4.6.2   Requirements of Testing IEC 61730-2

The test sequence from the second edition of IEC 61730-2: 2016 is shown in Figure 4.7. This, and the subsequent descriptions of the test sequence, are the authors interpretation of the IEC 61730-2 test sequence designed to familiarize the readers with the tests, not to provide alternative methods for performing the tests. Anyone actually performing the IEC 61730-2 safety test sequence should use the procedures as published in the IEC 61730-2 document.

In this version, all of the tests have been given MST numbers so these will be used in the explanation of module flow and then in the descriptions of the test procedures. This edition of IEC 61730-2 requires nine modules plus one unframed module as well as an additional module if you wish to qualify to PD 1. Additional samples are also needed if the module type is to be qualified with cemented joints. Requirements for these samples will be discussed in the sections on those tests – the Peel Test (MST 35) or the Lap Shear (MST 36).

The nine modules and one unframed module plus the extra one for qualifying to PD 1 are used in the following ways:

- All nine modules, the unframed module and the optional PD 1 module go through Visual Inspection (MST 01).
- One module is then pulled out as a control module. Its performance at STC is measured (MST 02) and it is then stored until being used as the control through the final set of measurements.
- A second module is pulled out after the initial Visual Inspection and subjected to the Module Breakage Test (MST 32). Nothing more is done with this module after the Breakage Test.
- A third module is pulled out after the initial Visual Inspection and subjected to the Ignitability Test (MST 24). Remember, that this test is not required in the US under UL 61730-2.
- The six remaining modules, the unframed module and the optional PD 1 module are then subjected to Maximum Power Determination (MST 03) and the Insulation Test (MST 04).
- At this point, the unframed module goes through the Impulse Voltage Test (MST 14) and a final Visual Inspection (MST 01). Nothing more is done with the unframed module.

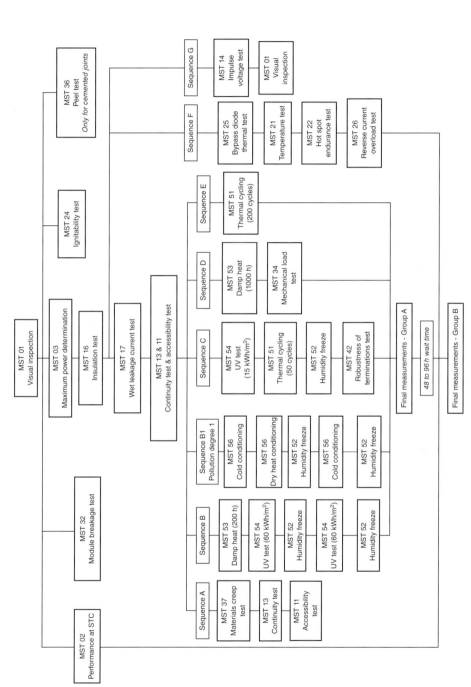

Figure 4.7 Accelerated stress test sequence from the second edition of IEC 61730-2: 2016. *Source*: redrawn by the Author for clarification.

- The six remaining modules and the optional pollution PD 1 module are subjected to the Wet Insulation Test (MST 17).
- One module is then pulled from the batch and subjected to Sequence F which includes the Bypass Diode Thermal Test (MST 25), the Temperature Test (MST 21), the Hot Spot Test (MST 22) and the Reverse Current Overload Test (MST 26). After each of these four tests, the module is evaluated using Visual Inspection (MST 01), the Insulation Test (MST 16) and the Wet Leakage Current Test (MST 17). After all of these tests, this module rejoins the others to go through Group B of the final measurements.
- The remaining five modules plus the optional PD 1 module then go through the Continuity Test (MST 13) and the Accessibility Test (MST 11). These modules are then split up into individual test sequences.
- One module goes through Sequence A which includes the Materials Creep Test (MST 37) followed by evaluation using Visual Inspection (MST 01), the Insulation Test (MST 16) and the Wet Leakage Current Test (MST 17). Then, that module is subjected to the Continuity Test (MST 13) and the Accessibility Test (MST 11). Nothing more is done with this module.
- One module goes through Sequence B which included 200 hours of Damp Heat (MST 53), 60 Kwh/m^2 of UV (MST 54) on the front side of the module, 10 Humidity-Freeze Cycles (MST 52), 60 Kwh/m^2 of UV (MST 54) on the rear side of the module and finally a second 10 Humidity-Freeze Cycles (MST 52). After each of these tests, the module is evaluated using Visual Inspection (MST 01) and the Insulation Test (MST 16). At the end of all of those tests, the module is evaluated using the Wet Leakage Current test (MST 17). Then it rejoins the others to go through Group A of the final measurements.
- One module goes through Sequence C which includes 15 Kwh/m^2 of UV (MST 54), 50 Thermal Cycles (MST 54), 10 Humidity-Freeze Cycles (MST 52) and Robustness of Termination (MST 42). After each of these tests, the module is evaluated using Visual Inspection (MST 01) and the Insulation Test (MST 16). After the Humidity Freeze and again after Robustness of Termination the module is evaluated using the Wet Leakage Current Test (MST 17). After this sequence, this module rejoins the others to go through Group A of the final measurements.
- One module goes through Sequence D which includes 1000 hours of Damp Heat (MST 53) and Mechanical Load (MST 34). After each of these tests the module is evaluated using Visual Inspection (MST 01), the Insulation Test (MST 16) and the Wet Leakage Current Test (MST 17). After all of these tests, this module rejoins the others to go through Group A of the final measurements.
- One module goes through Sequence E which includes 200 Thermal Cycles (MST 51). After this test, the module is evaluated using Visual Inspection (MST 01), the Insulation Test (MST 16) and the Wet Leakage Current Test (MST 17). After all of these tests, this module rejoins the others to go through Group A of the final measurements.
- The optional module for PD 1 goes through Sequence B1 which includes Cold Conditioning (MST 55), Dry Heat Conditioning (MST 56), 10 cycles of Humidity Freeze (MST 52), a second exposure to Cold Conditioning (MST 55) and a second 10 cycles of Humidity Freeze (MST 52). After each of these tests, the module is evaluated using Visual Inspection (MST 01) and the Insulation Test (MST 16). After the second

Humidity Freeze, the module is evaluated using the Wet Leakage Current Test (MST 17). After this sequence, the module rejoins the others to go through Group A of the final measurements.

- The modules that have gone through Sequences B, C, D and E and the optional module for PD 1 that went through Sequence B1, are all subjected to Group A of the final measurements. Group A begins with the Cut Susceptibility Test (MST 12) followed by evaluation using Visual Inspection (MST 01), the Insulation Test (MST 16) and the Wet Leakage Current Test (MST 17). Then the modules go through the Continuity Test (MST 13) and the Accessibility test (MST 11). After these two tests, they are joined by the Control Module and the module from Sequence F for final characterization in Group B of the final measurements.
- Group B of the final measurements consists of Maximum Power Determination (MST 03), Visual Inspection (MST 01), Bypass Diode Functionality Test (MST 07), and the Screw Connections Test (MST 33). The module from Sequence B also goes through the Insulation Thickness Test (MST 04), which is not required in the US under UL 61730.

The criteria for a module type to successfully pass the requirements of the second edition of IEC 61730-2 are, that all test samples must meet all of the pass criteria of each individual test and that none of the modules lose electrical continuity during the testing in sequences A through F. The product does not comply with this standard if any module fails one or more of the tests. Please note that there is no requirement for the module to retain any amount of output power (PV performance) after completion of the tests. So, the module does not have to continue to perform as a PV module, it just has to be safe to pass IEC 61730.

The following subsections describe the tests specified in IEC 61730-2:

4.6.2.1 MST 01 – Visual Inspection
The Visual Inspection test procedure is identical to MQT 01 of IEC 61215-2 with the addition of several safety-related items to inspect. These include checking:

- The compliance of the markings with Section 5.2 of IEC 61730-1: 2016 after performance of the Durability of Markings Test (MST 05).
- For the presence of sharp edges in accordance with the Sharp Edge Test (MST 06).
- The minimum distances (creepage and clearance distances) as defined in Tables 3 and 4 of IEC 61730-1: 2016.

In IEC 61730-2, the Visual Inspection Test has its own list of what is considered a Major Visual Defect and, therefore, grounds for failure of the test. The safety list includes the defects found in IEC 61215-1 and adds bubbles and delaminations that can lead to conductive paths through insulators, markings not complying with MST 05 and edges not complying with MST 06.

4.6.2.2 MST 02 – Performance at STC
The Performance at STC is identical to MQT 06.1 in IEC 61215-2. It is performed after the initial stabilization MQT 19.1 of IEC 61215-2.

4.6.2.3 MST 03 – Maximum Power Determination

The Maximum Power determination is identical to MQT 02 of IEC 61215-2.

4.6.2.4 MST 04 – Insulation Thickness Test

The Insulation Thickness Test is to verify that the insulation within the module continues to meet the minimum insulation thickness for thin layers specified as a function of rated voltage in either Table 3 or Table 4 of IEC 61730-1: 2016 depending on the PV module's Class. Once again, remember that this test is not required in the US under UL 61730. The test is performed on polymeric materials used as front-side or back-side insulation. This test is not required for glass.

The procedure for the test calls for selecting three worst-case locations, those places where you would expect the insulation to be the thinnest. These could be over solder bond bumps, at the edges of the module or at indents in the insulator caused by processing. The standard does not provide a method for measuring the insulation layers, it just tells you to do it. The method chosen can be either destructive or non-destructive. Any method can be chosen as long as the measurement uncertainty is less than 10% of the actual thickness.

The pass criteria require that the measured values minus their uncertainty must be greater than the minimum values given in Tables 3 and 4 of IEC 61730-1: 2016.

4.6.2.5 MST 05 – Durability of Markings Test

The purpose of the Durability of Marking Test is to verify that the markings on the module, things like the name plate and any warning signs are legible and durable. Legibility is checked by Visual Inspection (MST 01). Durability is checked by rubbing the markings by hand using medium pressure for 15 seconds with a piece of cloth soaked with water and again for 15 seconds with a piece of cloth soaked with petroleum spirits. After this test, the marking must still be legible and show no curling. In addition, the marking plate(s) should not be easy to remove.

4.6.2.6 MST 06 – Sharp Edge Test

The purpose of the Sharp Edge Test is to verify that the accessible PV module surfaces are smooth and free of sharp edges or burrs so they do not damage the insulation of conductors coming in contact with them or pose a risk of injury to those handing and/or installing the modules. Compliance is checked by visual inspection (MST 01), but it offers an option of performing the sharp edge test described in ISO 8124-1 to confirm compliance.

4.6.2.7 MST 07 – Bypass Diode Functionality Test

The Bypass Diode Functionality Test is identical to MQT 18.2 in IEC 61215-2.

4.6.2.8 MST 11 – Accessibility Test

The purpose of the Accessibility Test is to demonstrate that a person cannot touch high-voltage live parts within a PV module. The two pieces of equipment used are a test figure (a cylindrical test fixture Type 11 according to Figure 7 of IEC 61032: 1997 [53]) and an ohmmeter or continuity tester. The ohmmeter or continuity tester is attached to the PV module's short-circuited terminals and to the test finger. All connectors, covers and plugs that can be removed by hand are – that is without using any tools. The test fixture is then probed around the module trying to touch the active circuit. The ohmmeter or continuity tester is monitored during the probing to see if electrical contact is made between the probe and the electric circuit of the module.

The module passes the Accessibility Test as long as the probe never touches any active live parts and the resistance between the module's active circuit and the probe is always equal to or greater than $1\,M\Omega$.

4.6.2.9   MST 12 – Cut Susceptibility Test

The purpose of the Cut Susceptibility Test is to determine whether polymeric front- and backsheets can withstand the routine handling of installation and maintenance. To determine this, a steel hacksaw blade with a defined radius of curvature is dragged over the frontsheet or backsheet using a wheeled cart. The test is repeated five times in different directions, including over areas that look like they have circuit elements such as interconnects or bus bars underneath.

After the test, you repeat the Visual Inspection (MST 01), the Insulation Test (MST 16) and the Wet Leakage Current Test (MST 17). The module must pass each of these subsequent tests to pass the Cut Susceptibility Test.

4.6.2.10   MST 13 – Continuity Test of Equipotential Bonding

The purpose of the Continuity Test is to verify that there is a continuous electric path between accessible conductive parts that are in direct contact with each other (e.g. parts of a metallic frame). A current equal to 2.5 times the maximum overcurrent protection rating of the PV module under test (a value specified by the module manufacturer) is passed between the manufacturer's designated point for equipotential bonding (or grounding) and an adjacent (connected) exposed conductive component that is the greatest distance from the grounding point. The applied current and the resultant voltage are measured. This measurement is repeated for all accessible metal parts and for all connections, terminals, and/or wires included or specified by the manufacturer for grounding the PV module. The module passes the Continuity Test if all of the measured resistances are less than $0.1\,\Omega$.

4.6.2.11   MST 14 – Impulse Voltage Test

The purpose of the Impulse Voltage Test is to verify that the insulation of the PV module is capable of withstanding high voltages from the utility lines or even nearby lightning strikes. The test equipment and shape of the high-voltage pulse are given in IEC 60060-1 [54].

During the Impulse Voltage Test, the module is covered in a conductive foil using a conductive adhesive. The foil is connected to the negative terminal and the PV leads to the positive terminal of the voltage generator. Three pulses as defined in IEC 60060-1 are applied to the module at the voltage levels defined in IEC 61730-1 and given in Table 4.10. The polarity is then reversed and three more pulses are applied to the module at the same voltage.

These are very high-test voltages. For the typical Class II module rated at 1000 V, the test is performed at 12000 V. A module passes the Impulse Voltage test if there is no evidence of dielectric breakdown or surface tracking of the PV module during the test and if it can pass the Visual Inspection (MST 01) and the Insulation Test (MST 16) after exposure to the impulse voltage.

Table 4.10 Impulse voltage test levels from IEC 61730-2.

Module rated Voltage (V)	Test Voltage for Basic Insulation (kV)	Test Voltage for Double Insulation (kV)
50	0.8	1.5
100	1.5	2.5
150	2.5	4.0
300	4.0	6.0
600	6.0	8.0
1000	8.0	12.0
1500	10.0	16.0

4.6.2.12 MST 16 – Insulation Test

The Insulation Test itself is identical to MQT 03 of IEC 61215-2, but the test voltages depend on the Class and maximum system voltage rating of the product under test.

For Class II the test voltage is 2000 V plus four times the maximum system voltage.

For Class 0 the test voltage is 1000 V plus twice the maximum systems voltage. This is the level specified in IEC 61215-2.

For Class III the voltage is always 500 V.

For testing cemented joints, the test voltage selected above must be multiplied by 1.35.

4.6.2.13 MST 17 – Wet Leakage Current Test

The Wet Leakage Current Test is identical to MQT 15 in IEC 61215-2.

For testing cemented joints, the test voltage must be multiplied by 1.35.

4.6.2.14 MST 21 – Temperature Test

The purpose of the temperature test is to determine the maximum reference temperatures for various components and materials used in the PV module. It is a characterization test, not a stress test.

The module under test is mounted onto a black painted platform extending 60 cm beyond the PV module on all sides. The module is biased near peak power using either a resistive load or a maximum power point tracker. Data is then measured for the following parameters:

- Environmental or air temperature in a range from 25 to 45 °C.
- Irradiance in a range from 700 to 1000 W/m^2.
- Wind speed less than 1 m/s.
- Measure temperature of
  - Frontsheet
  - Backsheet
  - J-box
  - Field Wiring Terminals
  - Insulation of Wiring Terminals

- Connectors
- Bypass Diodes.

Stabilized temperature data for each of these locations should be determined. This data is then normalized to an ambient air temperature of 40 °C to give the value of the Temperature Test (T_{con}) using Eq. (4.4).

$$T_{con} = T_{obs} + \left(40°C - T_{AIR}\right) \tag{4.4}$$

Where T_{obs} is the measured temperature of the component and T_{AIR} is the temperature of the air or environment near the module during the measurements.

The Temperature values are then reported for the different components. The IEC 61730-2 standard says to pass the test "No measured temperatures exceed any of the applicable temperature limits (e.g. TI/RTE/RTI) of surfaces, materials, or components." However, that is really backwards of the way this test is used. A module manufacturer should do the Temperature Test first to determine what TI/RTE/RTI is necessary for that component or material and then, if necessary, replace the materials in the module with materials that have TI/RTE/RTI values that are higher than the measured T_{con}.

After the Temperature Test, the module must pass the Visual Inspection (MST 01), the Insulation Resistance (MST 16) and the Wet Leakage Current Test (MST 17).

4.6.2.15 MST 22 – Hot Spot Endurance Test
The Hot Spot Endurance Test is identical to MQT 09 from IEC 61215-2. Technology-specific requirements are given in the Technology Specific Sections, IEC 61215-1-x.

4.6.2.16 MST 24 – Ignitability Test
The purpose of this test is to determine whether impingement of a small flame on the surface or edge of the module will cause it to ignite and burn. Remember that in the US, under UL 61730, the Ignitability Test (MST 24) is not required because the US has its own required fire tests for modules.

The test is based on ISO 11925-2: 2010 [55]. The procedure allows for use of either a specially built test sample using the same construction materials as the module or a module itself. The test sample or module is oriented vertically during the test.

A specified flame from a gas burner is applied to the sample both on exposed surfaces at least 40 mm above the bottom edge of the specimen and on the edge. If the edge is protected by a non-flammable material like a metal frame the edge portion of the test may be omitted.

The flame is applied for a total of 15 seconds. The sample passes the test if it does not catch fire or if the flame does spread more than 150 mm vertically from the point of application of the test flame within 20 seconds from the time of initial application.

4.6.2.17 MST 25 – Bypass Diode Thermal Test
The Bypass Diode Thermal Test is identical to MQT 18 in IEC 61215-2.

4.6.2.18 MST 26 – Reverse Current Overload Test
The purpose of the Reverse Current Overload Test is to validate that the module can survive being exposed to a reverse current, equal to 135% of the PV module's overcurrent protection rating, which is equivalent to the module's series fuse rating. The PV module under

test is placed with its front face down on a single layer of white tissue paper on top of a support that holds the module but is a poor thermal conductor so that it does not conduct away any heat generated by the module. The back surface of the PV module is then completely covered with a single layer of white tissue paper in contact with the module. The cells should be in the dark (irradiance of less than $50\,W/m^2$) which can be achieved either by the covering or by performing the test in a darkened laboratory.

A DC power supply is used to provide current to the PV module. A reverse current equal to 135% of the PV module's overcurrent protection rating (module's series fuse rating), as specified by the manufacturer is applied to the module. The power supply should be in the current limited mode with the test voltage increased to achieve the desired reverse current through the module. The voltage should be adjusted to keep the current stable with ±2% during the course of the test. The test is continued for two hours, or until a failure has occurred.

The module fails the Reverse Current Overload Test if it, or the white tissue paper, show any signs of flaming or charring. After the Reverse Current Overload Test the module must pass the Visual Inspection (MST 01), the Insulation Resistance (MST 16) and the Wet Leakage Current Test (MST 17).

4.6.2.19 MST 32 – Mechanical Breakage Test

The purpose of the Module Breakage Test is to demonstrate that a module mounted as instructed by the manufacturer will not break in a dangerous manner when struck by a defined load. This test is based on ANSI Z97.1 [56].

In the Module Breakage Test, the test module is mounted vertically using one of the manufacturer's recommended method(s) onto a defined framing system that is bolted to the floor of the building so it will not move. A punching bag type impactor is filled with 45.5 kg of weights and is suspended on a cable within 13 mm of the center of the module. The impactor is then lifted to a height of 300 mm from the surface of the PV test module and then dropped so that it strikes the test module.

The PV module successfully passes the module breakage test if it meets criteria: (a) and either (b) or (c):

a) The PV module may not separate from the mounting structure or from the framing.
b) No breakage occurs.
c) If breakage of the PV module occurs, no opening large enough for a 76 mm diameter sphere to pass freely through can develop and no particles larger than $65\,cm^2$ can be ejected from the sample. In order to allow measurement of the particle size, a soft surface should be placed on the floor below the test module.

If the module type is required to pass the Equipotential Bonding Test (MST 13) – that is, if it has exposed metal parts – then it should be subjected to and pass MST 13 before and after the Module Breakage Test.

4.6.2.20 MST 33 – Screw Connections Test – Test for General Screw Connections MST 33a

The purpose of the general screw connection test is to demonstrate that screws can be tightened multiple times without causing damage. This test requires that screws and nut that are either transmitting contact pressure or are likely to be tightened by the user, be tested by being tightened and loosened five times. IEC 61730-2 provides a table that defines the torque to be applied to the screw depending upon the type and thread diameter of the screw under test.

The screws pass the test, if there is no damage impairing use of the screw or its connection. After the test, it must still be possible to introduce a screw or nut made of insulating material in the intended manner.

4.6.2.21 MST 33 – Screw Connections Test – Test for Locking Screws MST 33b

The purpose of the test for locking screws is to validate that a device that locks threads upon heating produces an adequate connection. (These can only be used for connections that are not subject to torque during normal operation.)

A defined torque is applied to the screw – 2.5 Nm for thread size ≤ M 10 and 5.0 Nm for thread sizes > M 10. These locking screws pass if they do not loosen during the test.

4.6.2.22 MST 34 – Static Mechanical Load Test

The Static Mechanical Load Test is identical to MQT 16 in IEC 61215-2. The Wet Leakage Current Test (MQT 15) does not have to be conducted after MST 34.

4.6.2.23 MST 35 – Peel Test

The purpose of the Peel Test is to determine whether two components (consisting of a flexible and a rigid component or two flexible components) are adequately bonded together to be rated as cemented joints. This test is only performed when the manufacturer is trying to rate their products using the cemented joint approach to reduce the required creepage and clearance distances. The Peel Test is not applicable to rigid-to-rigid bonded assemblies. Those are evaluated using the Lap Shear Strength Test (MST 36). The methodology of the Peel Test is taken from ISO 813 [57].

If the cemented joint to be tested is >10 mm in width, two standard laminates (one as a control and the other exposed to the tests in Sequence B) can be used for the Peel Test. If the cemented joint is ≤10 mm in width, special samples must be fabricated using a release sheet between the encapsulant and the backsheet to eliminate the encapsulant/backsheet adhesion from the measurement. Once again, one sample is a control and the other goes through the tests in Sequence B.

The Peel test is performed using a Tensile Testing machine that is capable of moving at least 50 mm/minute and has a fixture that grips the backsheet (or frontsheet) to pull at 90° ± 10° from the plane of the bond.

Ten strips (5 per interface) 10 mm ± 0.5 mm wide and at least 100 mm long are cut at the flexible frontsheet or flexible backsheet of the samples as shown in Figure 4.8. Five strips are to be cut per adhesion interface, five on the front and five on the back of the joint. The strips are to be cut from the same side of the module, but to different depths to reach the appropriate adhesion interface. That is cutting through the layer to be tested but not into any layers below.

The test pieces are placed symmetrically in the tensile testing fixture. The free end of the strip is placed in the grip. The tensile-testing machine is moved at 50 ± 5 mm/min until separation of the layers is complete. The force is recorded as a function of the movement.

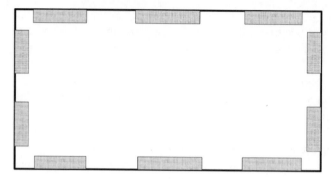

Figure 4.8 Layout of ten peel test strips for peel test.

This measurement is used to determine what force is required to cause separation of the layers. A plot of time versus force is created over the full length of the test piece. The adhesion strength in newton per mm is determined by dividing the measured force (in N) by the width of the test piece (in mm). Only samples that show a continuous peel-off characteristic for at least 20 mm can be used for subsequent calculations. Even if the measured maximum force deviates significantly from the continuous force, the continuous force is used in the calculations.

The mean value of adhesion is calculated for the five strips from the module that went through Sequence B and for the five strips from the control module. The module type passes the Peel Test if the average adhesion after Sequence B is greater than 50% of the value measured for the control module. The measurement can be validated with as few as three good values of adhesion from each module. If fewer than three are available additional strips should be cut and pulled.

4.6.2.24 MST 36 – Lap Shear Strength Test

The purpose of the Lap Shear Strength Test is to determine whether two rigid-to-rigid adhered components are adequately bonded together to be rated as cemented joints. This test is only performed when the manufacturer is trying to rate their products using the cemented joint approach to reduce the required creepage and clearance distances. The Lap Shear Test is not applicable to flexible-to-rigid or flexible-to-flexible bonded assemblies, which are evaluated using the Peel Test (MST 35). The methodology of the Lap Shear Test is taken from ISO 4587 [58].

Samples are built using two pieces of glass, one that is the same as the front glass and one the same as the back glass of the modules, but they need not be tempered. The glass samples should be 100 mm long and 25 mm wide. The cemented joint adhesive is applied to the end of one of the two pieces of glass in an area 12.5 mm by 25 mm. The two pieces are then bonded together as shown in Figure 4.9. Twenty such samples are fabricated. Ten are subjected to Sequence B tests while another 10 are used as controls. Then the adhesion of all 20 samples is measured using the tensile tester.

From the results, the mean value of adhesion is calculated for the 10 samples that went through Sequence B and for the 10 samples that were not stressed. The module type passes the Lap Shear Test if the adhesion after Sequence B is greater than 50% of the value from the control samples.

Front glass

← Adhesive

Back glass

Figure 4.9 Construction of lap shear samples for lap shear strength test.

4.6.2.25 MST 37 – Materials Creep Test

The purpose of the material creep test is to determine whether the materials used in the PV module will creep (that is move or flow) or lose adhesion when the module operates at the highest temperatures it normally experiences in the field. The test consists of placing the module in the worst-case mounting recommended in the installation instructions inside a temperature chamber. Vertical would typically be the worst case for creep if that is permitted in the installation instructions. The module is then heated to $105 \pm 5\,°C$ or $90 \pm 5\,°C$ if the module is designed for use only in open-rack systems (not roof mounted or insulated). The module is maintained at this temperature for 200 hours.

After the Creep Test, the module must successfully pass the Visual Inspection (MST 01), the Accessibility Test (MST 11), the Continuity Test (MST 13), the Insulation Test (MST 16) and the Wet Leakage Current Test (MST 17).

4.6.2.26 MST 42 – Robustness of Termination Test

The Robustness of Termination Test is identical to MQT 14 in IEC 61215-2. Both parts of MQT 14 (14.1 and 14.2) should be performed in that order. Junction boxes that have already been qualified to IEC 62790 must still be subjected to the tests in MQT 14.2. The Wet Leakage Current test (MQT 15) does not have to be conducted after MST 42.

4.6.2.27 MST 51 – Thermal Cycling Test

The Thermal Cycling Test is identical to MQT 11 in IEC 61215-2. Figure 4.7 shows that 50 thermal cycles are used in test sequence B and 200 thermal cycles are used in test sequence E. The Technology Specific sections of IEC 61215-1-x define the current flows used during the thermal cycles. The Wet Leakage Current test (MQT 15) does not have to be conducted after MST 51.

4.6.2.28 MST 52 – Humidity Freeze Test

The Humidity Freeze Test is identical to MQT 12 in IEC 61215-2. The Wet Leakage Current test (MQT 15) does not have to be conducted after MST 52.

4.6.2.29 MST 53 – Damp Heat Test

The Damp Heat Test procedure is identical to MQT 13 in IEC 61215-2. Figure 4.7 shows that a 200 hour damp heat test is used in test sequence B, while a 1000 hour damp heat test is used in test sequence D (the same duration as in IEC 61215-2). The Wet Leakage Current test (MQT 15) does not have to be conducted after MST 53.

4.6.2.30 MST 54 – UV Test

The process in this UV Test is equivalent to MQT 10 in IEC 61215-2 except that in IEC 61730-2, different UV dose levels are used. In Sequence C, the standard UV dose as described in IEC 61215-2 ($15\,kWh/m^2$) is used. In Sequence B, a UV dose equal to four times the IEC 61215-2 dose ($60\,kWh/m^2$) is applied twice, once to the front side of the module and once to the backside of the module. The Wet Leakage Current Test (MST 17) need not be performed after the UV tests in IEC 61730-2.

4.6.2.31 MST 55 – Cold Conditioning

The Cold Conditioning Test is part of Sequence B1 that is used to qualify modules to PD 1. The test is very simple. The module is cooled to a temperature of $-40\pm3\,°C$ and held there for 48 hours. After the Cold Conditioning test, the module must pass the Visual Inspection (MST 01) and the Insulation Resistance Test (MST 16).

4.6.2.32 MST 56 – Dry Heat Conditioning

The Dry Heat Conditioning is also part of Sequence B1 that is used to qualify modules to PD 1. The test is also very simple. The module is heated to $105\pm5\,°C$ with humidity below 50% and held under these conditions for 200 hours. If the PV module type under assessment is designed for use only in open-rack deployment the temperature for this test may be reduced to $90\pm3\,°C$. After the Dry Heat Conditioning Test, the module must pass the Visual Inspection (MST 01) and the Insulation Resistance Test (MST 16).

4.6.2.33 Recommendations for Testing of PV Modules from Production

IEC 61730-2 provides recommendations for production level testing. Since they are recommendations, they are not required. The recommendations include:

- Measurement of module output power on 100% of all modules.
- Wet Insulation Resistance on at least one module per laminator per shift. However, for frameless modules and for modules with cemented joints it recommends 100% testing.
- Visual Inspection on 100% of all modules to verify that clearance distances (distances of live parts to PV module edges) are within the product specification.
- Bypass Diode Functionality test on 100% of all modules.
- Continuity test on at least one module per framing table per shift.

Without performing these measurements, it is unclear how a module manufacturer would ensure the initial quality of its product.

UL 61730-2 (for use in the US) on the other hand, requires production level testing. Its requirements include the following measurements to be made on 100% of the production modules:

- Measurement of module output power.
- Either the Wet Insulation Resistance Test or the Insulation Test
- Visual Inspection to verify that clearance distances (distances of live parts to PV module edges) are within the product specification.
- Bypass Diode Functionality test
- Continuity test.

This set of measurements provides the necessary initial characterization of the modules to demonstrate that they meet the product performance and safety specifications.

References

1 Hoffman, A.R. and Ross, R.G. (1978). Environmental qualification testing of terrestrial solar cell modules. *Proceedings of the 13th IEEE PV Specialists Conference*, Washington, DC, USA (5–8 June 1978).

2 (a) IEC 61215 (2016). Terrestrial photovoltaic (PV) modules – Design qualification and type approval (Series).(b) IEC 61215-1 (2016). Part 1: Test requirements.(c) IEC 61215-1-1 (2016). Part 1–1: Special requirements for testing of crystalline silicon photovoltaic (PV) modules.(d) IEC 61215-1-2 (2016). Part 1–2: Special requirements for testing of thin-film Cadmium Telluride (CdTe) based photovoltaic (PV) modules.(e) IEC 61215-1-3 (2016). Part 1–3: Special requirements for testing of thin-film amorphous silicon based photovoltaic (PV) modules.(f) IEC 61215-1-4 (2016). Part 1–4: Special requirements for testing of thin-film Cu(In,GA) (S,Se)2 based photovoltaic (PV) modules.(g) IEC 61215-2 (2016). Part 2: Test procedures.

3 JPL Document No. 5101-162 (1981). Block V Solar Cell Module Design and Test Specification for Residential Applications.

4 Smokler, M.I., Otth, D.H., and Ross, R.G. (1985). The Block Program Approach to Photovoltaic Module Development. 18th IEEE PVSC in Nevada, USA (21–25 October 1985).

5 Jaffe, P. (1980). LSA Field Test Annual Report. JPL Document 5101-166, DOE/JPL-1 012-52, California, USA, Jet Propulsion Laboratory

6 Smokler, M.I. (1977). *User Handbook for Block II Silicon Solar Cell Modules*. JPL Document 5101–36. California, USA: Jet Propulsion Laboratory.

7 Smokler, M.I. (1979). *User Handbook for Block III Silicon Solar Cell Modules*. JPL Document 5101–82. California, USA: Jet Propulsion Laboratory.

8 Ross, R.G. and Smokler, M.I. (1986). *Flat-Plate Solar Array Project. Volume 6: Engineering Sciences and Reliability*. JPL Document 86–31. California, USA: Jet Propulsion Laboratory.

9 Smokler, M.I. (1982). *User Handbook for Block IV Silicon Solar Cell Modules*. JPL Publication 82–73, JPL Document 5101–214. California, USA: Jet Propulsion Laboratory.

10 Mon, G.R., Orehotsky, J., Ross, R.G., and Whitla, G. (1984). Predicating Electrochemical Breakdown in Terrestrial Photovoltaic Modules. 17th IEEE PVSC in Florida, USA (1–4 May 1984).

11 Ross, R.G. (1982). Photovoltaic array reliability optimization. *IEEE Trans. Reliab.* R-31 (3): 246.

12 UL 1703 (2002). Standard for Flat-Plate Photovoltaic Modules and Panels.

13 Wohlgemuth, J.H., Cunningham, D.W. Nguyen, A.M., and Miller, J. (2005). Long Term Reliability of PV Modules. 20th EUPVSEC in Barcelona, Spain (6–10 June 2005).

14 IEC 60904-9 (2007). Photovoltaic devices – Part 9: Solar simulator performance requirements.

15 IEC 60904-2 (2015). Photovoltaic devices – Part 2: Requirements for photovoltaic reference devices.

16 IEC 60904-1 (2006). Photovoltaic devices – Part 1: Measurement of photovoltaic current-voltage characteristics.

17 IEC 60891 (2009). Photovoltaic devices – Procedures for temperature and irradiance corrections to measured I-V characteristics.

18 IEC 61853-1 (2011). Photovoltaic (PV) module performance testing and energy rating – Part 1: Irradiance and temperature performance measurements and power rating.

19 IEC 61853-2 (2016). Photovoltaic (PV) module performance testing and energy rating – Part 2: Spectral responsivity, incidence angle and module operating temperature measurements.

20 IEC 60904-3 (2016). Photovoltaic devices – Part 3: Measurement principles for terrestrial photovoltaic (PV) solar devices with reference spectral irradiance data.

21 IEC 60904-7 (2008). Photovoltaic devices – Part 7: Computation of the spectral mismatch correction for measurements of photovoltaic devices.

22 Wohlgemuth, J. and Herrmann, W. (2005). Hot Spot Tests for Crystalline Silicon Modules. 31st IEEE PVSC in Florida, USA (3–7 January 2005).

23 Silverman, T.J., Mansfield, L., Repins, I., and Kurtz, S. (2016). Damage in monolithic thin-film photovoltaic modules due to partial shade. *IEEE J. Photovolt.* 6 (5): 1333–1338.

24 IEC 62790 (2014). Junction boxes for photovoltaic modules – Safety requirements and tests.

25 IEC 62938 Publication expected in 2020 Non-uniform snow load testing for photovoltaic modules.

26 Whipple, M. (1993). The Performance of PV Systems. NREL/DOE Photovoltaic Performance and Reliability Workshop, Colorado, USA.

27 Hibberd, B. (2011). PV Reliability and Performance: A Developer's Experience. 2011 NREL PVMRW.

28 Wohlgemuth, J.H., Cunningham, D.W., Nguyen, A.M., and Miller, J. (2005). Long Term Reliability of PV Modules. 20th EU PVSEC in Barcelona, Spain (6–10 June 2005).

29 Wohlgemuth, J.H., Cunningham, D.W., Amin, D. et al. (2008). Using accelerated tests and field data to predict module reliability and lifetime. 23rd EU PVSEC in Valencia, Spain (1–5 September 2008).

30 Wohlgemuth, J.H. (1993). Testing for module warranties. NREL/DOE Photovoltaic Performance and Reliability Workshop, Colorado, USA

31 (a) IEC 61730-1 (2016). Photovoltaic (PV) module safety qualification – Part 1: Requirements for construction.(b) IEC 61730–2 (2016). Photovoltaic (PV) module safety qualification – Part 2: Requirements for testing.

32 IEC 62930 (2017). Electric cables for photovoltaic systems with a voltage rating of 1.5 kV DC.

33 IEC 62852 (2014). Connectors for DC-application in photovoltaic systems – Safety requirements and tests.

34 IEC 62790 (2014). Junction boxes for photovoltaic modules – Safety requirements and tests.

35 IEC 62788-2 (2017). Measurement procedures for materials used in photovoltaic modules – Part 2: Polymeric materials – Frontsheets and backsheets.

36 IEC 60112:2003&AMD1 (2009). Method for the determination of the proof and the comparative tracking indices of solid insulating materials.

37 IEC 60216-5 (2008). Electrical insulating materials – Thermal endurance properties – Part 5: Determination of relative thermal endurance index (RTE) of an insulating material.

38 UL 746B (2018). Polymeric Material – Long Term Property Evaluations.

39 IEC 60216-1 (2013). Electrical insulating materials – Thermal endurance properties – Part 1: Ageing procedures and evaluation of test results.

40 IEC TS 62788-7-2 (2017). Measurement procedures for materials used in photovoltaic modules – Part 7-2: Environmental exposures – Accelerated weathering tests of polymeric materials.

41 IEC 60695-11-10 (2013). Fire hazard testing – Part 11-10: Test flames – 50 W horizontal and vertical flame test methods.

42 IEC 60695-10-2 (2014). Fire hazard testing – Part 10-2: Abnormal heat – Ball pressure test method.

43 UL 790 (2004). Standard Test Methods for Fire Tests of Roof Coverings.

44 IEC 60950-1 (2005). Information technology equipment – Safety – Part 1: General requirements.

45 ISO 1456 (2009). Metallic and other inorganic coatings – Electrodeposited coatings of nickel, nickel plus chromium, copper plus nickel and of copper plus nickel plus chromium.

46 ISO 1461 (2009). Hot dip galvanized coatings on fabricated iron and steel articles – Specifications and test methods.

47 ISO 2081 (2018). Metallic and other inorganic coatings – Electroplated coatings of zinc with supplementary treatments on iron or steel.

48 ISO 2093 (1986). Electroplated coatings of tin – Specification and test methods.

49 IEC 61140 (2016). Protection against electric shock – Common aspects for installation and equipment.

50 IEC 60664-1 (2007). Insulation coordination for equipment within low-voltage systems – Part 1: Principles, requirements and tests.

51 IEC 60529 1989+AMD1:1999+AMD2: (2013). Degrees of protection provided by enclosures (IP code).

52 IEC 62788-1-2 (2016). Measurement procedures for materials used in photovoltaic modules – Part 1-2: Encapsulants – Measurement of volume resistivity of photovoltaic encapsulants and other polymeric materials.

53 IEC 61032 (1997). Protection of persons and equipment by enclosures – Probes for verification.

54 IEC 60060-1 (2010). High-voltage test techniques – Part 1: General definitions and test requirements.

55 ISO 11925-2 (2010). Reaction to fire tests – Ignitability of products subjected to direct impingement of flame – Part 2: Single-flame source test.

56 ANSI Z97.1 (2015) Safety glazing materials used in buildings – safety performance specifications and methods of test.

57 ISO 813 (2016). Rubber, vulcanized or thermoplastic – Determination of adhesion to a rigid substrate – 90 degree peel method.

58 ISO 4587 (2003). Adhesives – Determination of tensile lap-shear strength of rigid-to-rigid bonded assemblies.

5

Failure Analysis Tools

Regardless of whether a module has degraded from field exposure or accelerated stress testing, it is important to understand what has actually changed within the module that led to lost peak power. If we want to use the results to improve the module construction so that future modules will not degrade, we must understand what particular changes have occurred. In this chapter, we will explore some of the methods used to better understand what has gone wrong within the module. Methods reviewed include, analysis of the I–V parameters, measurement of performance at different irradiances, visual inspection, Infrared (IR) Inspection, Electroluminescence (EL) and evaluation of adhesion. Each will be discussed in the subsections that follow.

5.1 PV Performance – Analysis of Light I–V Curves

Just knowing that the peak power has gone down doesn't tell us why. By analyzing the data from the I–V curve we can often get a better idea of what has degraded. Figure 5.1 shows an I–V curve from a fairly typical PV module. Three points are marked on the curve:

- I_{sc} is the short-circuit current of the module or the point at which the module voltage drops to zero.
- V_{oc} is the open-circuit voltage of the module or the point at which the current drops to zero.
- P_{max} is the maximum power point, which is the point on the I–V curve where the product of current and voltage is highest. This is where you would like the module to operate.

One additional parameter of importance is the fill factor (FF). It is defined as the maximum power divided by the product of the short-circuit current and open-circuit voltage as shown in Eq. 5.1.

$$FF = P_{max}/\left(I_{sc} * V_{oc}\right)$$ (5.1)

To evaluate why the P_{max} of a module has degraded it is of value to look at which of the three parameters I_{sc}, V_{oc} or FF is lower than expected.

Photovoltaic Module Reliability, First Edition. John H. Wohlgemuth.
© 2020 John Wiley & Sons Ltd. Published 2020 by John Wiley & Sons Ltd.

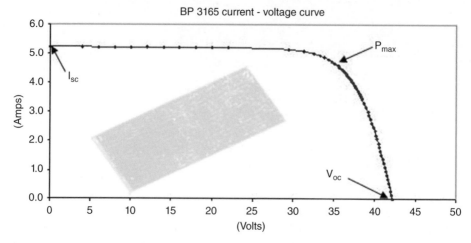

Figure 5.1 I–V curve of typical PV module. *Source:* prepared by Author for BP Solar commercial literature.

Reduction in short-circuit current, I_{sc} can be a result of one of the following issues:

- Reduction in the amount of light transmitted into the solar cells. This can be due to degradation of encapsulant which usually appears as a yellowing or darkening of the material as shown in Figure 2.10. It can also be due to a degradation of one of the anti-reflective (AR) coatings, either the AR coating on the glass or the AR coating on the cells. When the AR coating on the glass degrades, it usually becomes splotchy and irregular. When the AR coating on the cell degrades, it usually changes color from the typical dark blue often getting lighter, but not uniformly. Finally, soiling can also reduce the amount of light reaching the cells. Most soiling is easy to identify visually and is also the easiest factor to alleviate by cleaning the front surface.
- Loss of surface passivation. Many cells utilize surface passivation on the front of the cell to increase current collection. If this is the cause of the current reduction, it will not be visible to the naked eye like many of the other problems discussed in this section. Since loss of front-surface passivation will affect the light absorbed closet to the front surface, one signature of this degradation mode will be a reduction in the blue or short wavelength response of the module, which may be observed in the spectral response curve of the module.
- Loss of cell area via cracking (cry–Si) or corrosion (TF). Since the short-circuit current is proportional to the area of the cells that are connected in series, a loss of area for one cell leads to that cell not being able to produce as much current as the other cells. It doesn't take too much loss in one cell's area before it begins to impact the measured short-circuit current of the whole module. For crystalline silicon cells, this is most likely to occur when a cell breaks and an area becomes detached from the rest of the cell as shown by the dark area identified by the arrow in Figure 5.2. For thin-film modules, it is more likely that a part of one or more cells are corroded as shown in the lower right-hand side of Figure 5.3 and therefore do not produce as much current.

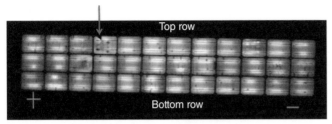

EL image of exposed module (837753)

Figure 5.2 Electroluminescence (EL) picture of cell with a crack that removes area from circuit. *Source:* reprinted from [1] with permission of Govindasamy Tamizhmani of Arizona State University.

Figure 5.3 Corrosion in thin film cells that removes active cell area.

Reduction in open-circuit voltage, V_{oc} can be a result of any of the following causes:

- Open circuit within a cell string – This results in the loss of voltage from all of the cells protected by the bypass diode that will turn on because of the open circuit. Since each diode usually protects multiple cells this means a large drop in voltage. If the module has four diodes, then a single open circuit within the cell string will result in loss of 25% of the module's open-circuit voltage.
- Bypass diode shorted – This provides a conductive path around the cells and will also lead to a significant reduction in voltage. However, unlike the previous case, the reduction will be minimized at open-circuit voltage when there is no current flowing through the module.
- Cell junctions shunted – Figure 2.30 showed an EL picture of some cells shunted by Potential-Induced Degradation (PID). All of the black cells in this picture are shunted so not producing the voltage they should be, therefore, the whole module will have lower voltage.

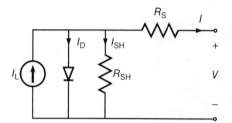

Figure 5.4 Equivalent circuit of a solar cell. *Source:* reprinted from [2] via permission granted by the copyright holder Squirmymcphee as provided on the website.

- Loss of surface passivation – Many types of solar cells have surface passivation that is designed to reduce surface recombination. Such surface recombination reduction is used to increase the open-circuit voltage in solar cells. A reduction in surface passivation, however, will likely only reduce the voltage for each cell by a few percent. So this type of degradation can be distinguished from some of the other causes of voltage degradation.

Before looking at what reductions in fill factor (FF) imply, it is best to look at the equivalent circuit of a module (or solar cell) to better understand what parameters are involved in determining the FF. This equivalent circuit in Figure 5.4 is based on a one-diode model for a solar cell or module.

From this equivalent circuit, an equation (Eq. 5.2) can be written for the Voltage (V) and the current (I) that appears at the output terminal of the solar cell or module.

$$I = I_L - I_o * \left\{ \exp\left[q\left(V + I * R_s\right) / \left(nkT\right)\right] - 1\right\} - \left\{\left(V + I * R_s\right) / R_{SH}\right\} \tag{5.2}$$

Where:

I is the current at the solar cell or module terminals (The I in the I–V curve)
V is the voltage at the solar cell or module terminals (The V in the I–V curve)
I_L is the photo-generated current usually assumed to equal the short circuit current I_{SC}
I_o is the reverse saturation current of the diode
n is the diode ideality factor, which would be equal to one for an ideal diode
q is the electronic charge
k is Boltzmann's Constant
T is the temperature of the junction in degrees K
kT/q is called the thermal voltage and at room temperature is equal to 0.0259 V
R_s is the Series resistance of the cell or module
R_{SH} is the shunt resistance of the cell or module

As we can see, there are three circuit elements that reduce the output of the solar cell or module and impact the FF.

- Series Resistance (R_s) – Current flow through the series resistance results in a voltage drop and therefore loss of output power. The higher the series resistance the more power is lost. Series resistance can be due to (i) the bulk resistance of the semiconductors themselves, (ii) the bulk resistance of the metallic contacts and/or Transparent Conductive Oxide (TCO), (iii) the contact resistance between the metal grid or TCO and the semiconductor, (iv) the resistance between the cell metallization and the interconnect ribbons,

(v) the bulk resistance of the interconnect ribbons, bus bars and the output leads and finally (vi) the contact resistance between any of the conductors like interconnect ribbons to bus bars or output leads to bus bars.

- Shunt Resistance (R_{sh}) – Any current that flows through the shunt resistance doesn't make it to the terminals of the solar cell and is therefore a loss of output power. The lower the shunt resistance the more power is lost. Shunt resistance can be due to PN junction leakage, leakage around the edges of the junction and foreign impurities and crystal defects in the semiconductor near the junction.

- Diode leakage (I_D) – Any current that flows through the leakage diode also doesn't make it to the terminals of the solar cell and is therefore a loss of output power. This leakage current is represented by the reverse saturation current of this diode (I_o) and by a diode ideality factor (n). The higher the leakage current and the higher the ideality factor, the more power is lost. This impacts both the open circuit voltage of the cell, and the shape of the I–V curve. It is best seen using the dark I–V curve and so will be discussed in Section 5.3. Diode leakage current is usually due to a poor junction in the solar cell where there are impurity levels near the junction that trap charge.

So, if a low FF (or a weak knee in the I–V curve) is observed, it could be due to a problem with series resistance or shunt resistance. Knowing which one is responsible will help to determine what has gone wrong with this particular cell or module. Since the equation is very complex, it is easiest to start out by assuming that the series and shunt resistances do not cause any power loss. This means a zero series resistance and an infinite shunt resistance. The top curves in both Figure 5.5a and Figure 5.5b are calculated with these assumptions. In Figure 5.5a we then add more and more series resistance to draw the subsequent I–V curves. In Figure 5.5b we have slowly lowered the shunt resistance for each subsequent curve to show its impact on the I–V curves. Please remember that these are just simplified models used to demonstrate the impact of series and shunt resistances on cell or module behavior.

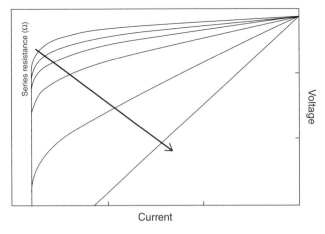

Figure 5.5a Effect of series resistance on I—V curve, arrow is in the direction of increasing series resistance.

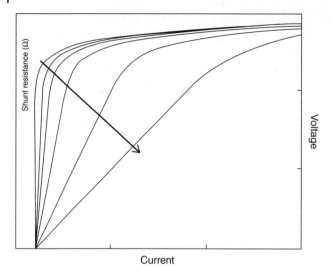

Figure 5.5b Effect of shunt resistance on I–V curve, arrow is in the direction of decreasing shunt resistance.

Figure 5.5a shows a set of I–V curves with increasing series resistance. Higher series resistance has the greatest effect on the portion of the curve toward the higher voltage end. As the series resistance grows larger, the curve tends to follow a straight line from V_{oc} to an intersection with the I_{sc} value at a non-zero voltage.

Figure 5.5b shows a set of I–V curves with decreasing shunt resistance. Lower shunt resistance affects the lower voltage portion of the curve near I_{sc} first. As the shunt resistance reduces, the curve tends to follow a straight line from I_{sc} to higher voltages creating a new knee in the curve at lower power.

So, if your I–V curve looks like one of these plots you may be able to tell whether your cell has too high a series resistance or too low a shunt resistance. However, it is often hard to tell whether a cell or module is suffering from series or shunt resistance from a single I–V curve, especially if the degradation is only a few per cent. If this is the case, it may be better to use the method described in the next section, taking I–V curves at a variety of irradiance levels.

5.2 Performance as a Function of Irradiance

As indicated in the previous section, it is often hard to tell what is happening in a module from the one sun (STC) I–V curve. Taking measurements at different irradiance levels as described in IEC 61853-1 [3] can be very useful. The module or cell can be measured at specific irradiances (and temperatures) or monitored outdoors for an extended period of time and the data extracted. To illustrate the usefulness of this method, two examples will be given, one for three different types of thin films and one for two supposedly similar crystalline silicon technologies.

Three early thin-film module types were deployed at the Solar Energy Centre near New Delhi, India from 2002–2003 to 2011 [4]. All three module types (triple junction a-Si, CdTe and CIGS) suffered significant power loss although the triple junction a-Si still met the manufacturer's power specifications indicating that much of the power loss was probably due to the expected light-induced degradation. All three module types suffered significant degradation of FF. To better understand what was happening in these modules, they were sent to National Renewable Energy Laboratory (NREL) for further analysis.

NREL confirmed the degradation of the three types of thin-film modules [5]. To better understand their behavior, the performance of the triple junction a-Si, CdTe and CIGS modules were measured as a function of irradiance. The set of irradiance curves, one for each type of module, are shown in Figure 5.6a–5.6c. To help understand the irradiance dependence, efficiency versus irradiance has been plotted for all three of the module technologies in Figure 5.7.

The CIGS module has good efficiency at low irradiance, but this degrades at higher irradiance levels indicating that there is a series resistance problem. The CdTe module shows little efficiency dependence on irradiance. This is fairly typical of newer generation, non-degraded CdTe modules, so it is probably not an indication of why these modules have

Figure 5.6a I—V curves as a function of irradiance for one of the CIGS modules deployed in India for eight years [5]. *Source:* reprinted from author's PVSC article.

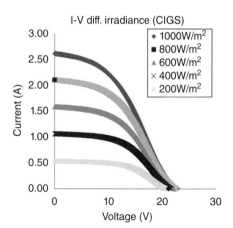

Figure 5.6b I—V curves as a function of irradiance for one of the CdTe modules deployed in India for eight years [5]. *Source:* reprinted from author's PVSC article.

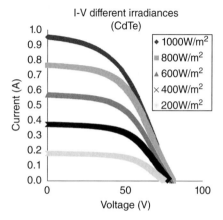

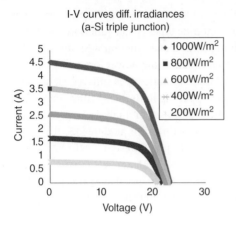

I-V curves diff. irradiances
(a-Si triple junction)

Figure 5.6c I—V curves as a function of irradiance for one of the triple junction a-Si modules deployed in India for eight years [5]. *Source:* reprinted from author's PVSC article.

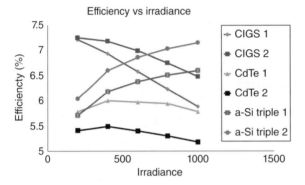

Figure 5.7 Efficiency as a function of irradiance for CIGS, CdTe and a-Si Triple Junction Modules [5]. *Source:* reprinted from author's PVSC article.

suffered power loss. Finally, the triple junction a-Si modules have their highest efficiency at one sun and gets worse at lower irradiances. This is contrary to the manufacturer's literature and field experience with similar modules in other systems. This poor low light level behavior appears to be due to the design of these particular modules, as the manufacturer has incorporated 22 bypass diodes into each module. The bypass diodes cause leakage around the cell junctions at low irradiance levels.

So, by looking at module performance as a function of irradiance we have determined that these CIGS modules suffered from an increase in series resistance and that these triple junction modules have poor low light level performance because of incorporation of too many bypass diodes.

The second example involves two solar arrays at NREL that were made up of multi-crystalline Si modules that were supposed to be very similar in technology and had similar one sun performance. However, the behavior of the two arrays was quite different with one array producing considerably more energy than the other. In order to understand why, two modules from each of the two arrays were evaluated. A useful parameter to look at in this case is Performance Ratio, defined as the ratio of the measured power at a specific irradiance divided by the STC power scaled for the irradiance as given in Eq. 5.3.

$$PR = \left(P_x / P_{STC} \right) / \left(G_x / G_{STC} \right) \tag{5.3}$$

Where P_x is the power measured at an irradiance of G_x.

The performance ratio shows how the module is performing at different irradiance levels during the day. Figure 5.8 shows a plot of the performance ratio for two modules from each of the two arrays plotted versus irradiance (Data for plots provided by Ryan Smith of NREL, *Personal Communication*). A high efficiency single-crystal control module has been added for comparison. The data shows that both modules from the B array (the one producing the lower amount of energy) have a lower performance ratio at the lower irradiance levels. The A-type modules have almost identical performance ratios that are greater than one in the mid-range of irradiances where series resistance is having less effect but shunt resistance hasn't started to reduce the efficiency. This is the shape of the performance ratio curve of many typical cry-Si modules. The single-crystal control does has significantly higher STC efficiency, but does not have as good a performance ratio as the type-A modules because it has a lower shunt resistance which results in efficiency loss at lower irradiances.

The separate impacts of series and shunt resistance are better seen by plotting FF versus irradiance as shown in Figure 5.9. In this plot, it is clear that the FF of the A modules improves as the irradiance goes down as the FF of this module type is dominated by series resistance. For B type modules, the FF starts increasing as the irradiance goes down, but

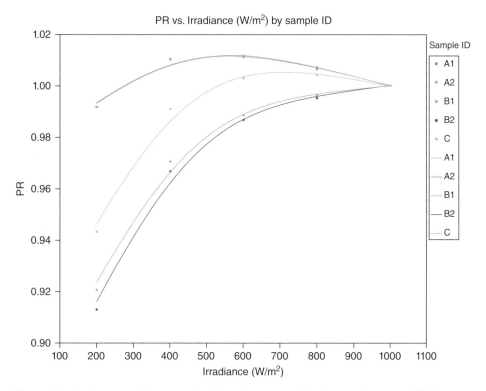

Figure 5.8 Performance Ratio versus Irradiance for two modules from each array at NREL.

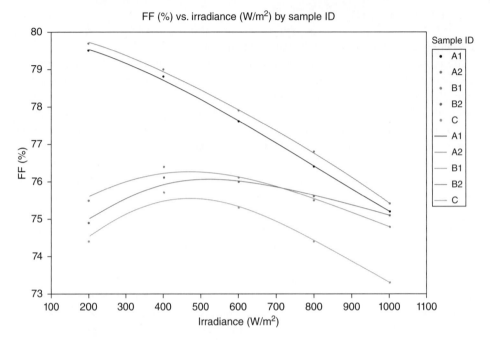

Figure 5.9 Fill Factor versus Irradiance for same module set.

this rolls over below about $500\,W/m^2$ as the poor shunt resistance starts decreasing the efficiency. In this plot, the single-crystal control looks more like the poorer performing Type B modules.

Another way to see the impacts of series and shunt resistance is to plot the dark I–V curve of the module. This will be discussed in the next section.

5.3 Dark I–V Curves

Dark I–V measurements can give us additional information about the parameters of solar cells or modules. To take a dark I–V curve an external power supply is used to push or pump current through the solar cells while they are in the dark. This current actually goes through the cells in the reverse direction from the light-induced current flow, so we must always remember that the current path is different in a dark I–V measurement than it is in a light I–V measurement. Because the basic relationship between voltage and current is an exponential, plotting the dark I–V curve on a linear graph will not provide much information. We need to plot the dark I–V curve on a semi-log plot with current on the logarithmic axis to see the relationships.

By looking at the various components of the diode equation (Eq. 5.2) we can identify how the different parameters impact the dark I–V curve. If the device is a perfect diode the curve will be a straight line. If the reverse saturation current (I_0 in Eq. 5.2) increases, the curve will retain its shape but move to the left (toward lower voltage for an equivalent current). If the diode ideality factor increases, the slope of the curve will change, becoming less steep.

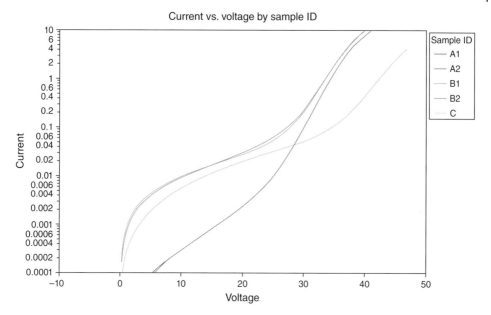

Figure 5.10 Dark I–V curves for same module set.

Indeed, for some solar cells and modules, the dark I–V curve has regions with distinctly different slopes. In this case, a two-diode model may be required to model the device.

Series resistance affects the dark I–V curve at higher current levels causing the curve to roll over to a less steep slope, but remaining a curved line to differentiate it from a changed ideality factor. Shunt resistance affects the dark I–V curve at lower current levels causing a flattening out of the curve.

Now let's take a look at the dark I–V curves from the cry-Si modules discussed in the previous section (See Figure 5.10). If we look at the upper part of the curves, we see that all four multi-crystalline modules have similar behavior which is not surprising since they all have similar STC performance. The single-crystal module has higher voltage but a significant part of that is because it has more solar cells. At lower voltages, the curves from type A and type B modules diverge. The lower shunt resistance of the type B modules leads to higher leakage current which in operation will reduce the output power at lower irradiances. Interestingly, the high efficiency single-crystal module looks very much like the B type multi-crystalline modules at lower voltages indicating that this module also has a lower performance ratio due to lower shunt resistance.

5.4 Visual Inspection

Visual inspection is one of the easiest ways to evaluate changes that appear in PV modules due to field exposure or accelerated stress tests. However, the critical question is whether the changes that are visible are important and related to any degradation in output power that is occurring in the module. For example, one of the easiest visual changes to observe is color change. It is not unusual to see color change listed as the most observed visual defect, but it is also one that is not likely to result in module failure, just some degradation in power over time. On the other hand, some of the failure modes that occur in modules have no (or minimal) impact on how the modules appear. For visual inspection to be a useful tool, it should be guided by experience.

The IEC 61215 Qualification Test protocol provides a method for performing visual inspections (See MQT 01 in Section 4.3) as well as a list of observed items that are considered major visual defects. However, this inspection is for single modules to be inspected in the laboratory. In many cases, we are interested in inspecting large numbers of modules in the field to provide information on the health of a particular PV array. Data regarding observations of module degradation in the field tend to vary in detail, consistency, and statistical significance (See as examples References 6, 7 and 8). NREL has led an attempt to standardize the collection of this data by developing a tool for the evaluation of visually observable defects in fielded PV modules [9]. A key goal in this project is to deliver a data collection tool for the comprehensive evaluation of visual defects that appear over time in PV modules.

The concept of data collection tool is to standardize what to look for and how to characterize the different observations so that results from one inspection can be compared to results from another. Because the results are consistent, they can then be evaluated statistically. The tool consists of instructions and data logging sheets with 14 sections based on the different module components, with inspection beginning at the rear side of the module and then continuing on to the front. Every attempt was made to adhere to IEC standard's terminology. There has been an attempt to balance the collection of sufficient detail for failure mode evaluation against the desire to minimize recording time per module. The forms themselves can be found along with the referenced report on the NREL website. The 14 sections will be discussed below. In most cases, the types of defects seen are chosen from a predefined list. In this way, the recorded data can be tabulated and used for statistical analysis.

1) Systems Data: The first section asks questions about the design and operation of the system including what type of system it is, single- or multi-module. If a multi-module system, it asks where in the system the module being inspected comes from. Then it asks how the system has been operated, for example, has it been peak-power tracked and finally whether the system has been grounded and if so how. Often some of these questions cannot be answered so they are just left blank.

2) Module Data: The next section asks questions about the particular module under inspection. Questions here include its technology, what certifications/qualifications it carries and an estimate of when it was deployed. Then it asks for details of its pedigree like who manufactured it and what are its model and serial numbers. Finally, it asks for the data on the nameplate like rated power, voltage etc. if this is still readable. If at all possible, it recommends taking a picture of both the front and the back of every module so you have a record and can go back and check if you missed observing or recording something later on.

The inspection then begins on the back side of the module.

3) <u>Rear side glass:</u> If the module has rear side glass you answer the questions in this section. If not, you go on to section 4. The rear side glass section asks if the rear side glass has any damage and if it does to identify if it is (i) crazed, (ii) shattered, (iii) cracked or (iv)chipped. It then asks how many cracks or chips are observed and approximately where they are located on the module.

4) <u>Backsheet:</u> If the module has a rear side polymeric backsheet you answer the questions in this section. If not, you skip it and go on to section 5. The backsheet section asks about the (i) appearance, (ii) texture, (iii) chalking and (iv) if there is any damage to the backsheet. Each of these gives the inspector choices to check. For appearance, the choices are (i) like new, (ii) minor discoloration or (iii) major discoloration. For texture, the choices are (i) like new, (ii) wavy but not delaminated, (iii) wavy and delaminated or (iv) dented. For chalking, the choices are (i) none, (ii) slight or (iii) substantial. If there is damage, the tool asks what type it is (i) burn marks (including selecting from a range of choices of the fraction of area burned), (ii) bubbles (including selecting the average bubble size (a) < 5 mm, (b) 5–30 mm or (c) > 30 mm as well as selecting whether the fraction of area that has bubbles is greater than 5 mm or not), (iii) delamination (including selecting the fraction of area delaminated and selecting the fraction of area where delamination has exposed parts of the circuit or cells), or (iv) cracks or scratches (including whether they are (i) random, (ii) over the cells or (iii) between cells and selecting the fraction of area affected by cracks or scratches and the fraction of area where they expose the electric circuit).

5) <u>Wires and connectors:</u> If the module has wires or connectors, or both, you answer the questions in this section. If not, you skip to section 6. This section asks for the condition of the wires and of the connectors, asking if they are (i) like new, (ii) pliable but degraded or (iii) embrittled. If the wire is degraded, it then asks if it is degraded by (i) cracks/disintegrating insulation, (ii) burning, (iii) corrosion, or (iv) animal bites. It then asks for identification of what type of connector it is. There are check boxes for some popular types but also boxes for "other" and "unsure." If the connectors are degraded, it then asks if they are degraded by (i) cracks/disintegrating insulation, (ii) burning or (iii) corrosion. Not sure what you do if your connector shows evidence of animal bites.

6) <u>Junction Boxes:</u> If the module has a junction box that is accessible you answer the questions in this section. If not, you skip to section 7. This section asks about the physical state of the junction box. Is it intact or is the structure unsound? If unsound it asks why; is it (i) weathered, (ii) cracked, (iii) burnt or (iv) warped? Then it asks about the condition of the junction box lid. Is it (i) intact, (ii) loose (iii) fallen off or (iv) cracked? Then it asks is there a junction box adhesive and if so is it observable? If not, you skip these questions. If it is observable it asks if the adhesive is (i) well attached, (ii) loose/brittle, or (iii) has the box fallen off? Is the adhesive itself (i) pliable, like new, (ii) pliable but degraded or (iii) embrittled? Attention then turns to the wire attachment in the box. If there is no wire or it is unobservable these questions are skipped. For modules with attached wires, evaluate how well attached the wires are to the junction box, the quality of the seal between the wires and the box (if there is no seal or the seal is compromised check "seal will leak"), and finally, if there is evidence of prior arcing.

7) <u>Frame Grounding:</u> The tool then asks whether the module has (i) a wired ground, (ii) a resistive ground, (iii) no ground or (iv) unknown. If there is a ground, the inspector then rates its appearance as (i) like new, (ii) some corrosion or (iii) major corrosion. Finally, there is a question about ground function – was the module well-grounded or was there no connection?

At this point, the inspector is asked to verify that they have taken pictures of the back of the module, the label and the junction box. Once that is verified the module is turned over to begin inspection of the front.

8) <u>Frame:</u> If the module has a frame you complete this section. If not, you skip this step and go on to section 9. The first question about the frame is whether it is (i) like new, (ii) damaged, or (iii) missing. If damaged, you determine if the frame has (i) minor corrosion, (ii) major corrosion, (iii) joint separation, (iv) cracking, (v) bending or (vi) discoloration, checking all that apply. Attention then turns to the frame adhesive selecting whether it is (i) like new or not visible or (ii) degraded. If it is degraded, you check all that apply from (i) adhesive oozed out or (ii) adhesive missing in areas.

9) <u>Frameless Edge Seal:</u> Once again, the inspector must decide whether the module has an edge seal or not. If not, you skip this step and go on to section 10. If it has an edge seal, you determine whether it is (i) like new, (ii) discolored (if so provide an estimate of the fraction affected with various ranges available for selection) or (iii) visibly degraded. Then, the inspector looks for material problems like (i) material squeezed or pinched out or (ii) material shows signs of moisture penetration. The final question in this section is about the fraction delaminated with various ranges available for selection.

10) <u>Glass/polymer front:</u> The first step in this section is to determine what the front sheet is made from (i) glass, (ii) polymer, (iii) glass/polymer composite or if it is (iv) unknown. Then the inspector is asked to determine if the front sheet is (i) smooth, (ii) slightly textured, (iii) pyramid/wave textured or if (iv) it has an anti-reflective coating. The next question is whether the front is (i) clean, (ii) slightly soiled or (iii) heavily soiled. If soiled, it than asks whether the soiling is localized (i)near the frame (in which case you identify which side – (a) left, (b) right, (c) top, (d) bottom or (e) all sides) or (ii) is it localized at an event like a bird dropping. The next question is about observation of any damage with possible observations of (i) none, (ii) small localized damage or (iii) extensive damage. If damage has been observed the type is recorded as (i) crazing or other non-cracking damage, (ii) shattered (tempered), (iii) shattered (untempered), (iv) cracked (including how many cracks are observed by selecting a range and where it appears the cracks start (a) corner, (b) edge, (c) cell, (d) junction box and e) foreign object impact), (v) chipped (including how many chips are observed by selecting a range and where chips occur- in (a) module corners or (b) module edges) and finally (vi) milky discoloration (including selecting a range of areas over which the discoloration occurs).

11) <u>Metallization:</u> The first question in the metallization category concerns gridlines or fingers and whether (i) there are none or barely observable so the category has to be skipped or (ii) applicable. If applicable are they are (i) like new, (ii) have light discoloration or (iii) have dark discoloration? If discolored, the correct range is selected for the fraction discolored. The second question concerns bus bars and whether (i) this

category is not applicable or not observable or (ii) applicable. If applicable, are they are (i) like new, (ii) have light discoloration or (iii) have dark discoloration? If discolored, the correct range is selected for the fraction discolored. It then asks whether the bus bars have any of the following, checking all that apply: (i) obvious corrosion, (ii) diffuse burn marks and (iii) are misaligned. The third question concerns cell interconnect ribbons and whether (i) this category is not applicable or not observable or (ii) applicable. If applicable are they are (i) like new, (ii) have light discoloration or (iii) have dark discoloration? If discolored, the correct range is selected for the fraction discolored. It then asks whether the cell interconnect ribbons have any of the following, checking all that apply: (i) obvious corrosion, (ii) diffuse burn marks and (iii) have breaks. The fourth question concerns string interconnects and whether (i) this category is not applicable or not observable or (ii) applicable. If applicable, are they are (i) like new, (ii) have light discoloration or (iii) have dark discoloration? If discolored, the correct range is selected for the fraction discolored. It then asks whether the string interconnects have any of the following, checking all that apply: (i) obvious corrosion, (ii) diffuse burn marks, (iii) have breaks or (iv) have arc tracks.

12) Silicon (mono or multi) modules: If the module has crystalline-silicon cells you answer the questions in this section. If not, you skip to section 13. This section starts out with questions about the module construction before asking about the condition of the module. The construction questions start by asking how many cells are in the module. Then it asks how many cells are in series, followed by how many strings are in parallel. Next, the inspector is instructed to enter the width and length of the solar cells. The next question deals with the distance between the cells and the frame of the module, offering two selections (i) >10 mm or (ii) <10 mm. The following question deals with the distance between the cells in a string, offering two selections (i) >1 mm or (ii) <1 mm. The next sections address the condition of the module. The first question in this section deals with cell discoloration, asking whether there is (i) none observed (like new), (ii) light discoloration or (iii) dark discoloration. It then asks for the number of cells that are discolored. Looking at just the discolored cells, you are to select the range that best represents the average percentage of area of those cells that are discolored. Then the inspector is to indicate where the discoloration is occurring, selecting all that apply from (i) module center, (ii) module edges, (iii) cell center, (iv) cell edges, (v) over gridlines, (vi) over bus bars, (vii) over tabbing ribbon, (viii) between cells, (ix) are some individual cells darker than others and (x) partial cell discoloration. The final question in the discoloration subsection is whether the discoloration near the junction box is the (i) same, (ii) more or (iii) less than that elsewhere in the module. The next category is damage where you can check 'none' and go on to the next category or check all of the types of damage observed from the list of (i) burn marks (with a subsection to select the range of number of cells observed with burn marks), (ii) cracking (identifying how many cells are observed to have cracks), (iii) evidence of moisture penetration, (iv) worm marks or snail trails (identifying how many cells are observed to have snail trails), and (v) foreign particles embedded. The final category in this section looks at delamination. The inspector selects from (i) none, (ii) from edges, (iii) uniform, (iv) corner(s), (v) near junction box, (vi) between cells (indicating what fraction of the delamination is between cells), (vii) over cells (indicating what fraction

of the delamination is over cells) and (viii) near cell or string interconnects. The final chore in this section is to identify the likely interface for any delamination: (i) glass, (ii) semiconductor, (iii) encapsulant, (iv) backsheet or (v) bus bar.

13) <u>Thin-Film Modules:</u> If it is a thin-film module you check (i) applicable and complete the section, if not, you check (ii) not applicable and go on to section 14. The first set of questions is about characterizing the module before asking about its condition. The construction questions start by asking how many cells are in the module. Then it asks how many cells are in series, followed by how many strings are in parallel. Next, the inspector is instructed to enter the width and length of the solar cells. The next question deals with the distance between the cells and the frame of the module, offering two selections (i) >10 mm or (ii) <10 mm. The questions then turn to the condition of the module starting with appearance, asking if it is (i) like new, (ii) have minor/light discoloration or (iii) have major/dark discoloration. Then you report on the type of discoloration checking all that apply from (i) spotted, (ii) haze (encapsulant browning) or (iii) other. Then you report on the location of any discoloration selecting all that apply from (i) overall (no pattern), (ii) module center, (iii) module edges, (iv) cell center, (v) cell edges and (vi) near cracks. The second category addressed is damage where you select from (i) none, (ii) small, localized or (iii) extensive. Then the type of damage is reported selecting from (i) burn marks, (ii) cracking, (iii) evidence of moisture ingress or (iv) embedded foreign particles. The third category addressed is delamination where you select from (i) none, (ii) small, localized or (iii) extensive. Then you report on the locations of any delaminations selecting all that apply from (i) from outer edges, (ii) uniform, (iii) corner(s), (iv) near junction box, (v) near bus bars and (vi) along scribe lines. Finally, the type of delamination is reported as (i) absorber, (ii) AR coating or (iii) other.

   At this point, the inspector is asked to verify that pictures have been taken of the front of the module.

14) <u>Electronic Records:</u> The last section provides space to record whether various electronic records have been taken and, if so, where those files are stored. If no electronic records are taken, "not applicable" should be marked and you are finished with the inspection of this module. Spaces are provided to provide the filenames for electronic records of digital photo files, I–V curve(s), and EL and IR images. Space is also provided for recording the measured irradiance and temperature and the sensors used for their respective measurements during taking of the I–V curve(s). If an I–V curve was measured, the inspector should evaluate the connector on the module using one of the provided options: (i) functions, (ii) no longer mates, or (iii) exposed, meaning that the connector does not properly seal the electrical connection. The inspector may also choose to perform a test of the bypass diodes. If this test is performed, the total number of diodes should be recorded, along with the number of diodes that are found not to be working because they are either (i) shorted or (ii) open circuited.

Though a detailed knowledge of a module conditions is often necessary to diagnose the cause of degradation or failure, it may not be practical to collect all of the detailed data requested in the long-form data sheet. In cases where large numbers of modules must be processed quickly (particularly for field inspection of grid-scale installations), collecting a subset of data can still provide meaningful, yet less comprehensive, statistics for failure analysis purposes. For this purpose, NREL also developed a short-form collection tool that follows the same structure and format as the long-form collection, while removing much of the detail about damage location and quantification of damage extent. As a result, select data fields from both the long- and short-form can be merged for statistical analysis.

Even if you don't use the NREL developed tool for your inspection, just being familiar with it can help someone doing inspections by indicating what sort of things to look for during your inspection.

5.5   Infrared (IR) Inspection

IR cameras can be used to look at the heat generated within a PV module or system. By looking at the longer wavelength light emitted by the modules or other system components the camera identifies temperature differences within the device. Generation of heat within the PV module or the PV system usually means a defect or flaw has occurred, either an increase in resistance to current flow or possibly a situation where part of the modules is no longer producing any current at all.

Initially, IR cameras were large bulky units that required liquid nitrogen cooling to work. Their use was restricted to laboratories or manufacturing facilities where the equipment could be maintained. In this mode, they were used as a tool to determine why certain modules were not working properly. Modules were biased using a dc power supply and current approaching the STC short-circuit current was pushed through the modules. The resultant IR picture showed where the current was flowing and areas that were hotter or cooler identified defects. Figure 5.11 is an example of an IR picture of a module that has lost a significant amount of its power after thermal cycling. The IR image shows where several cell interconnect ribbons are dark – that is, they are not carrying current while the other interconnects on the same cells are hotter because they are carrying twice the current. It turned out that, in this module, the interconnect ribbons were soldered too close to the edge of the cell and so several broke during thermal cycling resulting in power loss and showing up as dark regions in the IR picture.

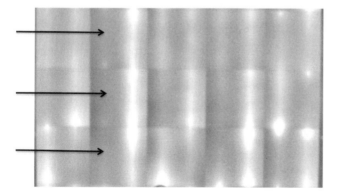

Figure 5.11   Infrared (IR) image of broken interconnect ribbons (see arrows).

Eventually, EL took over this function from IR cameras. (EL will be discussed in the next section). However, as IR cameras improved, they no longer needed liquid nitrogen cooling and were refined to the point where hand-held units were available. So, IR moved into the field. Today, IR cameras are mostly used to inspect operating PV systems in the field. In this mode, they can be used to look for hot spots in modules, bypass diode that are passing current when they shouldn't be and for any other circuit elements (such as connectors) that have degraded to the point where they are causing heating and probably loss of output power. One of the main advantages of IR in this case is its ability to take the images while the system is fully functioning, that is, while the sun is shining, so that the current path is the actual operating system current flow. Figure 5.12 shows an IR picture of a hot spot in a thin-film module. Figure 5.13 shows an optical picture of the same spot showing delamination of the bus tape at that point [10].

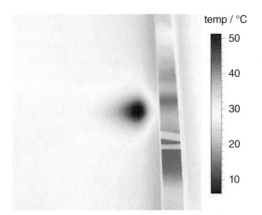

Figure 5.12 Infrared (IR) picture of hot spot in thin film module [10]. *Source:* reprinted from author's PVSC article with permission of Timothy Silverman of NREL.

Figure 5.13 Optical picture of bus tape delamination in thin film module [10]. *Source:* reprinted from author's PVSC article with permission of Timothy Silverman of NREL.

There are many other defects in a PV system that can be observed using IR imaging. There are now services available to have your operating PV system inspected using IR imaging. In some cases, the IR images are taken from drones or even airplanes. There is now also fairly extensive literature about using IR inspection of PV plants. Several useful papers that describe the use of IR cameras in PV can be found in References 11 and 12. In addition, IEC has published a new Technical Specification (IEC TS 62446-3) on outdoor infrared thermography [13].

5.6 Electroluminescence (EL)

In EL, the device emits light in response to an applied electric current. Light is emitted in a process called radiative recombination, which is the same principal that occurs in light emitting diodes (LEDs). Any defect within the PV device can disrupt the generation of the light and thus shows up in the EL signal. A variety of detectors including Si (CCDs) (Charged Coupled Devices) and InGaAs are used as EL detectors to evaluate PV cells and modules. Because the process requires detecting light (for silicon cells and modules it is near infrared light) EL is usually performed in the laboratory in the dark and not typically performed on operating arrays in the field. There have been some research efforts in trying to apply sinusoidal electric current and lock-in detection methods to look at operating arrays using EL but such approaches are still experimental and beyond the scope of this book.

In most EL measurements, a power supply is used to provide current flow through a PV cell or module in a darkened laboratory. A computer with a digital camera is used to record the resultant image. Not only are researchers using EL to see defects, but some are also performing quantitative analysis on the EL images. Reference 14 provides a good summary of EL technology applied to PV. IEC has recently published a Technical Specification (IEC 60904-13) that provides many of the details necessary to set-up and use an EL system to evaluate PV cells or modules [15]. Some of the defects that can be seen in PV modules (and cells) include:

Cracks: EL is the perfect tool for observing cracks in a PV module or cell. Figure 5.14 shows one of the earliest EL pictures taken at BP Solar which provided the first evidence that some of the manufacturing tools in one of the factories were putting too much pressure on the solder bonds and breaking the cells as evidenced by the Xs in the picture [16].

EL can also be used to distinguish between closed cracks that do not remove any area from the cell circuit (like those in Figure 5.14) and open cracks that remove active area from the cell circuit (like in Figure 5.15) where the dark regions indicate regions of the cell that are no longer producing any current.

Missing Grid Fingers: EL has such good resolution that it can often see missing grid fingers from the cells in a module. Figure 5.16 is an example of an EL picture that shows missing grid lines.

Shunted Cells: When a solar cell within a module is shunted (or shorted) it no longer produces any voltage and therefore has no EL signal. This often happens with PID degradation as was shown in Figure 2.30 where all of the cells in the left-hand row are dark – shunted cells producing no EL signal.

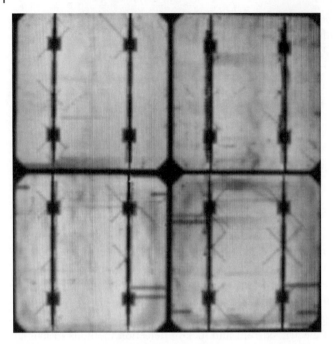

Figure 5.14   Electroluminescence (EL) picture of cells with cracks caused by manufacturing [16]. *Source:* reprinted from author's PVSC article.

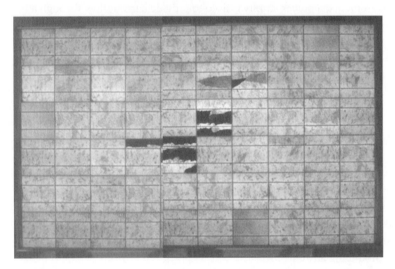

Figure 5.15   Electroluminescence (EL) picture of damaged cry-Si module. Black areas are disconnected from the rest of the module circuit and not producing an EL signal.

Silicon wafers with lower minority carrier lifetime: EL is so sensitive that it shows cells with lower minority carrier lifetimes, which appear darker than cells with higher minority carrier lifetimes. An example of this can be seen in Figure 5.17 where some of the cells are darker than others. At BP Solar, we were able to use EL to identify dark bands on the edges of cells made from wafers cut from the edge (one band) or the corner (two bands) of the Si ingot.

Figure 5.16 Electroluminescence (EL) picture showing cell with missing grid lines – the dark vertical lines in the picture. *Source:* picture provided by Govindasamy Tamizhmani of ASU.

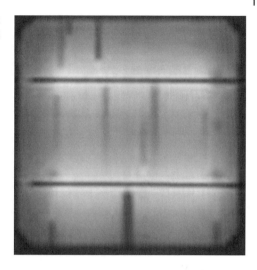

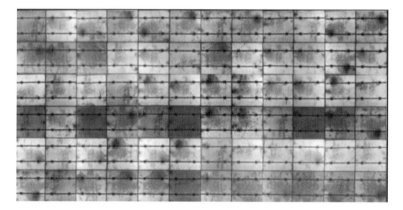

Figure 5.17 Electroluminescence (EL) picture of module showing some cells with lower minority carrier lifetimes (darker cells) than others (lighter cells) [16]. *Source:* reprinted from author's PVSC article.

Corrosion of grid lines: As discussed in Chapter 4, the Module Qualification test sequence contains a damp heat test where the modules are exposed to 85 °C at 85% relative humidity for 1000 hours. As the industry has begun to question whether the Qualification Tests are adequate for evaluating long-term performance (See Chapters 6, 7, 9 and 10) it is only logical that in research the various tests in the qualification sequence will be extended. When cry-Si modules are subjected to an extended damp heat test, they often begin to lose significant power after 2500–3000 hours of exposure. The reason for this can be seen by observing the EL image as a function of exposure time in Figure 5.18 [17]. After 3000 hours of damp heat exposure the outer grid lines on the cells are no longer actively collecting carriers. Their contact to the Si surface has been etched away. As will be discussed in the later chapters, this phenomenon is not being seen in the field so may be a test failure that does not duplicate a field failure.

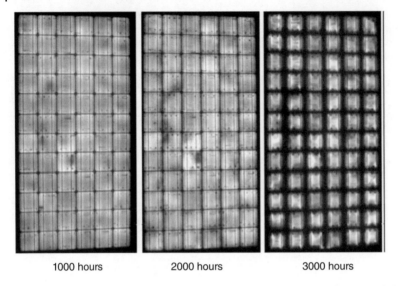

| 1000 hours | 2000 hours | 3000 hours |

Figure 5.18 Electroluminescence photographs of a cry-Si module after extended damp heat testing [18]. *Source:* reprinted from PVSC with the permission of the article's author, Werner Herrmann of TÜV Rheinland.

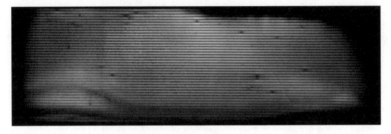

Figure 5.19 Electroluminescence (EL) picture of a thin film module after eight years of deployment in New Delhi, India [5]. *Source:* reprinted from author's PVSC article.

Degradation in thin-film modules: EL can also show a variety of degradation phenomena in thin-film modules. Figure 5.19 is one example of an early thin-film module built without edge seals that shows both shunting (the dark splotches throughout) and corrosion particularly around the edges at the right [5].

Since EL shows many of the degradation modes that occur in PV modules, it is very useful for evaluating modules after field exposure or accelerated stress testing. However, since EL shows so many different types of defects it can also be used as a QA tool for new modules. Many module manufacturers do 100% EL inspection of modules as part of their final inspection before shipment. Some bigger customers do 100% EL inspection of incoming modules before installation. In this case, there is some controversy over what constitutes a reject. For example, customers do not want to install modules with any cracked cells, while manufacturers claim that some cracks are benign and will never result in power loss. The jury is still out on this topic.

5.7   Adhesion of Layers, Boxes, Frames, etc.

Adhesion is a critical issue for long-term survival of PV modules. Of course, if the junction box or frame has come loose the mechanical and electrical integrity of the module is in jeopardy. Many of the other failure modes may also be the result of poor adhesion. For example, the corrosion of the cell metallization shown in Figure 2.7 was due to poor adhesion between the solar cell surface and the encapsulate (the EVA had no primer and while the glass was primed, the cells were not).

One of the main techniques for finding poor adhesion is via visual inspection. In the visual inspection tool, adhesion is an important category to document. Parts that are accessible like junction boxes and frames should be pulled on and, if possible, wiggled to assess whether they are still well adhered or not. For the laminate itself, you can look for voids, bubbles or any indication that the layers are coming apart, but it can be hard to tell which surface is visible or which materials have debonded. Use of low magnification can be useful in helping to identify which delaminated surface you are seeing. For example, the inside the glass often has texture that cannot be seen when the encapsulant is adhered to it, but it becomes visible when delaminated.

When there are no visible delaminations or voids, how do you determine whether the adhesion between different layers is still good or if it has deteriorated significantly? One tool for such an evaluation is a peel test. MQT 35 is such a peel test that is used in the module safety tests (IEC 61730), to evaluate whether two materials are well enough adhered to be considered a cemented joint. The advantage in this application is that the samples can be prepared with an accessible flap for initiating the peel. This cannot be done to evaluate the adhesion in samples after field exposure or accelerated stress testing. Starting the peel is difficult and leads to significant variability in the results making it hard to judge how much the adhesion has really changed during exposure or testing.

Another approach to evaluation of adhesion in PV modules is the use of a tensile test [19]. A button is glued to the back of one of the materials (for example, the encapsulant). The encapsulant (and any other materials in the stack) is scored around the edge of the button down to the surface whose adhesion is to be measured. A torque is then applied to the button until the surface adhesion fails and the button pops off. The torque required to remove the button is a measure of the adhesion.

While both the peel test and the tensile test can give a comparison of the adhesion on different areas of the same sample, they are not well suited to compare different types of samples. The measured results depend on additional factors including the elasticity of the materials being evaluated. A new method for measuring adhesion within PV modules is being developed under the PVQAT program and will be discussed in Chapter 7.

References

1 Tamizhmani, G. (2014). 20 Years Exposed Siemens M55 Modules at NREL: Test Results of I-V, EL, QE, Reflectance, UV Fluorescence and Infrared Imaging. Prepared for NREL.

2 Wikimedia (2015). https://commons.wikimedia.org/w/index.php?curid=5025465 (Accessed 28 August 2019).

3 IEC 61853-1 (2011). Photovoltaic (PV) module performance testing and energy rating – Part 1: Irradiance and temperature performance measurements and power rating.

4 Sastry, O.S., Singh, R.K., Chandel, R. et al. (2011). Degradation in Performance Ratio and Yields of Exposed Modules under Arid Conditions. 26th EU PVSEC in Hamburg, Germany (5–9 September 2011).

5 Wohlgemuth, J.H., Sastry, O.S., Stokes, A. et al. (2012). Characterization of Field Exposed Thin Film Modules. 38th IEEE PVSC in Texas, USA (3–8 June 2012).

6 Vazquez, M. and Rey-Stolle, I. (2008). Photovoltaic module reliability model based on field degradation studies. *Progress in Photovoltaics* 16: 419–433.

7 Dunlop, E.D. and Halton, D. (2006). The performance of crystalline silicon photovoltaic solar modules after 22 years of continuous outdoor exposure. *Progress in Photovoltaics* 14: 53–64.

8 Ishii, T., Takashima, T., Otani, K. et al. (Mar 2011). Long-term performance degradation of various kinds of photovoltaic modules under moderate climatic conditions. *Progress in Photovoltaics* 19: 170–179.

9 Packard, C.E., Wohlgemuth, J.H., and Kurtz, S.R. (2012). Development of a Visual Inspection Data Collection Tool for Evaluation of Fielded PV Module Condition. NREL/TP-5200-56154.

10 Wohlgemuth, J., Silverman, T., Miller, D.C. et al. (2015). Evaluation of PV Module Field Performance. 42nd IEEE PVSC in Colorado, USA (14–19 June 2015).

11 Moretón, R., Lorenzo, E., and Narvarte, L. (2015). Experimental observations on hot-spots and derived acceptance/rejection criteria. *Solar Energy* 118: 28–40.

12 King, D.L., Kratochvil, J.A., Quintana, M.A., and McMahon, T.J. (2000). Applications for infrared imaging equipment in photovoltaic cell, module, and system testing. 28th IEEE PVSC in Alaska, USA (15–22 September 2000).

13 IEC TS 62446-3 (2017). Photovoltaic (PV) systems – Requirements for testing, documentation and maintenance – Part 3: Photovoltaic modules and plants – Outdoor infrared thermography.

14 Guo, S., Schneller, E., Davis, K.O., and Schoenfeld, W.V. (2016). Quantitative Analysis of Crystalline Silicon Wafer PV Modules by Electroluminescence Imaging. 43rd IEEE PVSC in Oregon, USA (5–10 June 2016).

15 IEC 60904-13 (2018). Photovoltaic devices – Part 13: Electroluminescence of photovoltaic modules.

16 Wohlgemuth, J.H., Cunningham, D.W., Placer, N.V. et al. (2008). The Effect of Cell Thickness on Module Reliability. 33rd IEEE PVSC in California, USA (11–16 May 2008).

17 Wohlgemuth, J.H. and Kempe, M.D. (2013). Equating Damp Heat Testing with Field Failures of PV Modules. 39th IEEE PVSC in Florida, USA (16–21 June 2013).

18 Herrmann, W. and Bogdanski, N. (2011). Outdoor Weathering of PV Modules – Effects of Various Climates and Comparison with Accelerated Laboratory Testing. 37th IEEE PVSC in Washington, USA (19–24 June 2011).

19 Dhere, N.G. and Gadre, K.S. (1998). Tensile Testing of EVA in PV Modules. *Proceedings of the International Solar Energy Conference*, New Mexico, USA (14–17 June 1998).

6

Using Quality Management Systems to Manufacture PV Modules

The first five chapters of this book have presented the efforts involved in testing and evaluating the performance of PV modules. However, you cannot test quality into a product. Testing can only indicate whether a manufacturer has been successful at designing and building quality into the product in the first place. In order to continuously build quality modules, the manufacturers should be using Quality Management Systems that have been developed specifically for PV-module manufacturing.

This chapter will relate some of the history behind how Quality Management Systems evolved in PV, indicating how successful this has been as well as identifying some of the issues with the present system and the need for further improvements.

6.1 Quality Management Systems

A quality management system (QMS) is a collection of business processes and practices designed to continuously improve the quality of the products being produced to ensure that customer expectations and requirements are met or exceeded. A quality management system (QMS) is usually implemented as a framework of organizational structures, methods, procedures, techniques, policies, processes, and resource allocations established to provide the necessary control over all aspects of the company's operation. A QMS details the methods by which responsibilities and relationships are defined, ensures meeting schedules, and details how contracts and agreements are implemented.

So while a QMS impacts all aspects of the company's operations, in this book, we are most interested in how it impacts the quality of the products, particularly the PV modules being manufactured. So, let's look at a QMS in terms of manufacturing products. Quality assurance for products is a way of preventing mistakes and defects in manufactured products and avoiding problems when delivering solutions or services to customers. A QA plan for a product should contain the following components:

Specifications: The specifications contain a statement of what the product is being designed to do. For a PV module, it will contain a list of performance parameters, like Standard Test Conditions (STC) peak power, open-circuit voltage, short-circuit current and maximum-operating system voltage. It should also indicate what consensus standards the

Photovoltaic Module Reliability, First Edition. John H. Wohlgemuth.
© 2020 John Wiley & Sons Ltd. Published 2020 by John Wiley & Sons Ltd.

product meets. In the case of a PV module, this is likely to include module design qualification (IEC 61215) and module safety qualification (IEC 61730).

Design Process: This is the procedure used to determine what the product will look like and how it will perform, particularly in relation to the proposed specifications. The design process should be based on what the customers want so this process needs to have a marketing as well as engineering component. Part of this process must validate that the product can be manufactured with an acceptable cost. Design Qualification (for example IEC 61215 for PV modules) should also be an integral part of the design process.

Manufacturing Process: The manufacturing part of the QMS should provide the detailed instructions on how to fabricate the product including listings of all raw materials and components as well as all tooling/equipment required in the manufacturing process. This should be like a cookbook in that it describes everything necessary to manufacture the module. This means a step-by-step listing of the processes and any process controls necessary to ensure the products continue to meet the specifications.

The manufacturing process part of the QMS must be consistent with the equipment available, with the skill level of the employees on the production line and with the cost structure required for the product.

Raw Materials and Components: Manufacturing usually requires the purchase of raw materials and/or component parts. Specifications of those materials or components are often critical to the quality of the finished products. Many manufacturers qualify suppliers of these materials and components to help ensure both the quality and timely delivery of the procured material or component.

Inline Inspections and Measurements: Rather than wait until a product reaches final inspection, manufacturers utilize inline inspections and measurements to validate that the processes are in control. These measurements and inspections can be performed on a statistical sampling basis or on 100% of the in-process products. For example, during the manufacturing of PV solar cells, some manufacturers measure the sheet resistance on a sampling of the in-process wafers after diffusion to validate that the pn junction has been formed correctly. Similarly, in-process cry- Si solar cells are often inspected after screen printing to verify that the screen-printing process has produced the required grid finger width and height. Historically, a small sample of pull tests were performed on tabbed cells to verify that the soldering process was producing solder bonds with adequate strength.

Final Inspections and Testing: Once a product is complete, there are usually final inspections and tests or measurements that are made before the product can be shipped. For PV modules, this usually includes measurement of the STC peak power, short-circuit current and open-circuit voltage measured on 100% of the completed modules before shipment. There are also some safety-related measurements required by standards, particularly UL 1703 that require high pot testing on 100% of completed PV modules before shipment.

Non-conforming Products: A company should have a procedure for how it handles nonconforming products. When a problem or out-of-control process is discovered, it must be rectified. But what do you do with the products that have been made while the process was or might have been out of control? The QA plan should provide for isolation of that product and then a review of the consequences. Can the product be shipped as is, reworked or scrapped? Ultimately, the QA plan must indicate who will make the decision and what they will base that decision on. If there is enough product involved, the decision may require some accelerated stress testing to see if the product is susceptible to a particular failure mode.

Warranty and Returns: The product warranty should be part of the QMS. It will define what non-conformities in the product would authorize customers to return that product to the manufacturer for replacement or refund. The plan must then provide for a methodology for how returns are handled including assessing whether the returned products meet the definition in the warranty. There should then be a procedure to handle both cases, when the returned product meets the definition of defective from the warranty and when it does not.

As indicated in Chapter 4, Jet Propulsion Laboratories (JPL) recognized that PV module manufacturers needed QMS during the Block Buy program and helped those companies participating in the program to develop their own quality programs. So, during the formidable days of PV development many module manufactures utilized a QMS that they developed in-house.

6.2 Using ISO 9000 and IEC 61215

The Photovoltaic industry wasn't the only industry that was in need of guidance on what a good QMS should look like. The International Organization for Standardization (ISO) stepped into this gap and prepared a set of standards, the ISO 9000 series of standards, to provide the fundamental concepts and principles of Quality Management. The major goal of this series of standards was to help companies to improve their ability to consistently provide products and services conforming to their requirements. This set of standards was first published in 1987. Within the series of standards, ISO 9001 [1] defined the requirements for obtaining certification to ISO 9000.

From 1987 onward, manufacturers in many different industries began applying ISO 9000 to their organization and then having their QMS certified by an accredited agency for compliance. Hence, they could then call their factories certified to ISO 9000. This was a great advance for manufacturing because there was now a standard way to assess QMS. To meet ISO 9001, most manufacturers had to improve their QA systems and its documentation. The assumption being made that a product produced in an ISO 9000 certified factory is likely to have higher quality than a similar product produced in a factory without ISO 9000 certification.

The PV industry participated in this program starting in the 1990s with most, if not all, of the largest module manufacturers becoming ISO 9000 certified. The Quality norm in PV became a PV module qualified to IEC 61215 produced in an ISO 9000 certified factory.

The earliest versions of ISO 9000 (1987 and 1994) did not solve all of the manufacturing quality issues. ISO 9001 was intended to provide auditors with a consistent set of criteria with which to assess QMS. However, many in the industry applied ISO 9001 in an incorrect manner – focusing on conformity to ISO 9001 instead of focusing on their own quality management [2]. The idea propagated that ISO 9001: 1987 was intended for use by management to use as a guide for their QMS rather than as a guide to the auditors on how to assess a QMS. This meant that, in some cases, Quality Management Plans were written around the standard rather than around the production process. This can and, in some cases, did lead to problems.

Probably the worst issue was the notion that all the company had to do was define their QMS and then follow it and they would be ISO 9001 certified and be making quality products. What this leaves out is the question of whether the product being produced actually meets the customer's needs.

The ISO working group finally realized this and rectified many of these issues in the new ISO 9000: 2000 version. This update had five main goals.

1) Meet stakeholders needs
2) Be useable by all sizes of organizations (not just big companies)
3) Be useable by all business sectors
4) Be simpler and easier to understand
5) Connect Quality Management to the business process.

The 2000 version did a much better job of getting companies to design the right products for the application and then to tailor their QMS to their business processes.

There are a number of additional improvements in the 2015 version. It is based on seven principles.

1) Customer focus
2) Leadership
3) Engaging people
4) Process approach
5) Improvement
6) Evidence-based decisions
7) Relationship Management.

So, as the product standards have evolved so have QMS. For many years, buying a PV module with IEC 1215 made in an ISO 9001 certified factory was the best you could do.

6.3 Why just Using IEC 61215 and ISO 9000 is No Longer Considered Adequate?

This system of purchasing modules certified to IEC 61215 made in an ISO 9001 factory defined quality PV modules for many years through the 1990s and early 2000s. However, when customers started building really large PV arrays (multi-tens of Megawatts) they began to question whether this was adequate to guarantee the quality of the modules particularly in terms of long-term survival.

While the actual number of PV modules that failed (that is stopped producing any power) was quite low, everyone realized that PV modules do degrade over time, losing some of their output power. Dirk Jordon of the National Renewable Energy Laboratory (NREL) published a series of papers reviewing the degradation rates of PV modules and PV systems

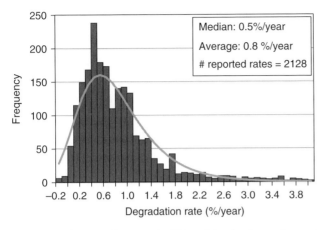

Figure 6.1 Degradation rates for PV modules (and arrays) reported in the literature [3]. *Source:* reprinted from *Progress in PV* with the permission of the publisher, Wiley, Inc. and the author, Dirk Jordan of NREL.

from the literature. Figure 6.1 shows a histogram of the reported degradation rates compiled from the literature [3]. The median degradation rate is 0.5% per year with an average of 0.8%. Typically, the module power warranty is based on no more than 20% power loss in 25 years of operation, which equates to a yearly power loss of less than 0.8%. From Figure 6.1, it can be seen that there is a significant tail of PV modules that have degraded more than this rate over time. Now there was data to show that there was product in the field at least some of which was qualified to IEC 61215 and had been built in an ISO 9000 certified factory that was not meeting the promised long-term performance level. So, what is happening here?

Why are we seeing such a significant percentage of the modules degrading in excess of 0.8% per year? There are several reasons for this. Remember in Section 4.5 (Limitations of the Qualification Tests) it was already stated that "Qualification tests are not designed to identify and quantify wear-out mechanisms." The tests used are of too short a duration to cause the modules to wear out or to differentiate between products that may have long and short lifetimes. So, the Qualification Tests are pretty good at identifying module types that would fail fairly quickly in the field, but not over the longer times we expect PV modules to survive.

In the earlier days of PV (the 1990s for example), there were only a handful of PV-module manufacturers including Solarex that merged with BP Solar, Arco Solar that became Siemens Solar that became Shell Solar, Sharp, Kyocera and Mobil Solar that became ASE Americas. These companies realized that the qualification tests were not sufficient for evaluation of long-term operations in the field and so each added additional accelerated stress tests to their QMS beyond the levels prescribed in the qualification tests to ensure longer module lifetimes.

As one example of such extended testing [4]. Solarex increased the number of thermal cycles from 200 to 400 when extending the warranty to 20 years and then increased it further to 500 cycles when the warranty increased to 25 years. The internal Solarex qualification process also increased the damp heat time from 1000 to 1250 hours for the 25-year warranty. These longer term tests went a long way to eliminating some of the design and process flaws that resulted in module degradation rates greater than the 0.8% per year level on which the warranties were based.

There are several additional reasons why we may be seeing an increase in the number of premature module failures associated with module types that have successfully passed the qualification test sequence.

- PV systems have changed significantly since the 1990s. Use of transformerless inverters results in ungrounded PV systems, which can lead to modules suffering from potential Induced degradation (PID) a failure mode virtually unknown 10 years ago and not tested for in the Qualification Test Sequence.
- In addition to use of ungrounded systems, industry is trending toward the use of ever higher system voltages which causes added stress on the PV modules. When the Qualification tests were developed most modules were being used in smaller, low-voltage systems.
- PV cells and modules are always changing. Larger, thinner cells are more prone to breakage. Larger modules suffer from higher mechanical stresses also increasing cell breakage. More broken cells are seen in today's modules after a few years of field deployment than were seen back in the 1990s when cry-Si cells were typically at least 300 µm thick.
- There is a continuing effort to reduce module cost. Sometimes, manufacturers cut corners that they shouldn't cut. While these changes are supposed to be requalified via re-testing according to the re-test guidelines in IEC 62915 [5], sometimes these tests are skipped altogether and sometimes the wrong tests are selected for re-test based on faulty engineering analysis of what is likely to go wrong because of the proposed change. In either case, premature module failures are possible.

By the mid-2000s, there were many new PV-module manufacturers entering the market. It is clear that many of these did little, if any, testing beyond the qualification tests before introducing their products to market. Govindasamy Tamizhmani at ASU published a paper documenting how the failure rates in the IEC 61215 qualification tests increased significantly in their test laboratory during the period 2005 to 2008 [6] as many new module manufacturers sought to qualify their modules.

6.4 Customer Defined "Do It Yourself" Quality Management and Qualification Systems (IEC 61215 on Steroids)

In response to reports of premature module failures and excessive degradation many PV system installers and purchasers began to develop and require their own set of extended accelerated stress tests. Usually, these were based on increasing the stress levels for the tests already in the IEC 61215 qualification test sequence. In 2012 at the NREL PV Module Reliability Workshop, a session was held in which different groups were invited to present their version of an extended accelerated stress test regime. In this session called "IEC 61215 on Steroids" [7] more than 10 different organizations presented their extended test protocols. Presentations were made by the following teams:

a) BP Solar, Q-Cells and VDE
b) Renewable Energy Test Center
c) NREL and Frauhofer Center for Sustainable Energy
d) TUV Rhineland
e) Atlas
f) Toray Industries
g) ESPEC Corp
h) Supsi – Swiss PV Module Test Centre
i) Florida Solar Energy Center
j) Underwriters Laboratories

No two of the test protocols were the same and none of the organizations were able to justify their tests by demonstrating that they identified field failures that are not identified by the qualification tests themselves.

At the same time that these groups were developing their own extended test protocols, some large purchasers of modules began to worry that module manufacturers were not using adequate quality management systems. They were basically saying that using ISO 9000 and being inspected to ISO 9000 criteria was not adequate. So, these module purchasers began requiring the factories to be inspected by an expert hired by the customer. Of course, because there was no standard for such inspections, these experts (like Solar Buyer) created their own set of criteria.

Therefore, to now obtain a large purchase order of modules, a module manufacturer could be required to have their modules tested to a different test sequence and their factory inspected against a different set of criteria. I think we can all see how this could lead to duplication of efforts and potentially serious problems.

6.5   Problems with the "Do It Yourself" System

There are two major problems with the "do it yourself" approach to product qualification. The first was indicated in the last section in that it may require module manufacturers to do more testing than is necessary. The second is that it may force module manufacturers to make costly but unnecessary changes to their products to pass tests that do not predict module performance.

As stated in the previous section, many of the new test sequences are based on increasing the test levels from IEC 61215. Sometimes this is useful as will be seen in the next Chapter where recent work indicates that 200 thermal cycles is not sufficient to test for 25 year module survival in many climates. However, in other cases, there doesn't appear to be any data that would support longer or more stressful exposures.

One of the tests often extended is the damp heat exposure test. To understand whether it makes sense to extend this test, let's review the test as it appears in the qualification sequence and look at some background information before evaluating the results of longer-term damp heat testing.

One of the things we must understand is that the 85 °C, 85% relative humidity condition used in the test is extreme. Figure 6.2 shows the modeled humidity levels and temperatures

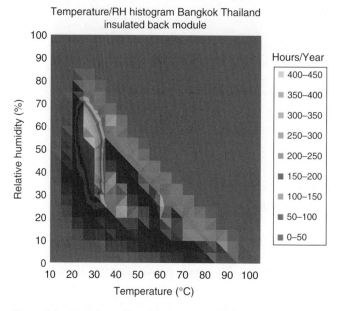

Figure 6.2 Modeling of humidity present within the back side of a glass/polymer module in Bangkok, Thailand [7]. *Source:* reprinted from the authors PVSC paper with permission of Michael Kempe of NREL.

within the back EVA of a glass/polymer module deployed in Bangkok, Thailand [8]. The test condition of 85/85 never occurs inside the module, even one deployed in one of the most humid climates in the world. As seen in the figure, when the module has high-relative humidity, it is fairly cool and when the module is hot, it has a low-relative humidity. With that said, the 85/85 test has been extremely successful at helping to eliminate module failures due to moisture ingress. We should remember that as practiced in IEC 61215 the damp heat test is already an extreme exposure.

Most PV module types now in the field are qualified to either IEC 61215 or IEC 61646, so they have successfully passed 1000 hours of damp heat testing at 85° C and 85% RH (85/85). As early as 2003, BP Solar extended its 85/85 testing to 1250 hours to internally qualify modules for a 25-year warranty [4]. A subsequent paper [9] reported on testing of 10 different cry-Si module types from nine international module manufacturers. All of these module types were qualified to IEC 61215, providing third-party documentation of passing 1000 hours at 85/85. However, in the BP Solar test, 8 of the 10 module types failed 1250 hours of 85/85 testing with >5% power loss. At that time, a slightly longer damp heat test, therefore, did differentiate module durability in a chamber. This is an indication that some manufacturers' module designs just barely pass the 1000-hour 85/85 test.

The philosophy of some testing laboratories appears to be that if 1000 hours of damp heat exposure is good, then 2000 or even 3000 must be better. Figure 6.3 shows the results from one of the longer-term experiments [10]. For the first 2000 hours of exposure, the module shows little power loss, but in the next ~1500 hours of 85/85 exposure, the module loses more than half of its efficiency. Subsequent reports have confirmed the tendency for EVA based screen-print modules to suffer significant performance loss after extended 85/85

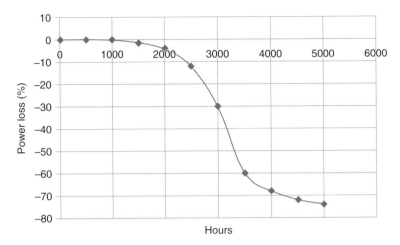

Figure 6.3 Long-term damp heat testing (85/85) results [9]. *Source:* reprinted from the author's PVSC paper.

testing. Several more recent publications have provided valuable insights into the mechanisms responsible for this power loss [11, 12]. Figure 5.18 showed the electroluminescence (EL) image for a module subjected to extended 85/85 exposure [11]. After 3000 hours of exposure to damp heat a dark area forms around the periphery of each cell. The dark area indicates that the periphery of each cell is no longer actively collecting carriers, but the grid lines themselves do not look corroded. It has been proposed that the moisture induces corrosion of the doped oxide that provides the electrical contact between the gridlines and the emitter-layer of the silicon cells [13]. As the time of the damp heat exposure increases, the size of the damaged region at the periphery grows, indicating that this moisture-induced process is penetrating further across the surface of the cells. Since current is collected from less cell area, the efficiency of the module decreases as shown in Figure 6.3.

Longer-term damp heat testing has also been applied to other PV technologies. Ketola of Dow Corning reported on long-term damp heat testing of cry-Si modules with a new silicone encapsulation system [14]. With this silicone encapsulation, the modules lost less than 5% of their power after 5000 hours of 85/85 testing. Tony Sample of ESTI reported on long-term damp heat testing of a variety of thin-film modules [15]. His work demonstrated that thin-film PV in well-sealed packages could survive beyond 1000 hours of 85/85 before suffering measureable degradation. There are also numerous reports that glass-on-glass cry-Si modules can survive at least 5000 hours of 85/85 without significant power loss [16]. It is really only screen-printed cry-Si cells encapsulated with EVA in glass/polymer backsheet packages that show this particular failure mechanism in damp heat testing. Is this something we really need to be worried about for long-term field survival?

There are only a limited number of field failure modes that have been reported in the literature (See Chapter 2 and Ref. [17]). From the list of failure modes, only a few are solely attributable to high-humidity levels including:

<u>Corrosion</u> of cells, cell metallization, interconnect ribbons, bus bars and connectors are all facilitated by moisture. It is seldom the semiconductor itself that corrodes but rather the

TCO or metallization. The native oxide on Si is sufficiently protective that we do not expect silicon cells to corrode. The extended failures in damp heat discussed above result from corrosion of the cell metallization, specifically the ohmic contact between the grid lines and the silicon. There were reported cases of infant mortality of modules resulting from corrosion of the cell metallization in the early days of PV. However, this corrosion had significantly different visual appearance than that seen from the extended damp heat testing. The field corrosion looks more like what was shown in Figure 2.7 where the metallization itself has clearly been corroded. When just the doped oxide corrodes, the grid lines themselves appear like new. Once modules began successfully passing the 1000-hour damp heat test and the humidity freeze test from IEC 61215, there has been little observation of such corrosion occurring in the field.

Delamination of the encapsulant from either the front glass or the solar cells is a failure mode seen regularly in the field. Many of the failures that appear to be related to moisture ingress occur in combination with delamination of the encapsulant (The example shown in Figure 2.7 is probably an example where the cell surfaces were not primed). In this case, the primary failure mode is corrosion of the front-side metallization of the solar cells facilitated by delamination of the encapsulant from the solar cells, leaving the cell metallization exposed to any liquid water that has condensed in the delamination pockets. The long-term damp heat tests do not typically result in delamination, but clearly can result in cell degradation. Delamination is not adequately evaluated in the present qualification test sequence and will be discussed further in Chapter 7, but adding more time to the damp heat test is not the right approach for evaluating delamination failures.

The type of power loss seen after long-term damp heat is not routinely reported as being observed after field exposure. Also, there does not appear to be any EL pictures from fielded modules that look anything like the one in Figure 5.18 taken after 3000 hours of damp heat exposure with dark rings around the periphery of each cell. To help understand why, let's take a look at some modeling work done by Mike Kempe at NREL. The failures seen in the extended damp heat test occur on the front of the cells. So, Kempe has modeled the ingress of moisture into the thin EVA layer located between the cells and the glass [18]. Figure 6.4

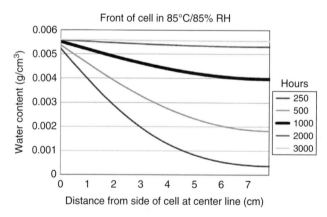

Figure 6.4 Results of analysis for the humidity level in the EVA located between the glass and a cell during damp heat testing. *Source:* printed with the permission of Michael Kempe of NREL.

shows the front-side moisture ingress for a 15.6 cm by 15.6 cm cell during 85/85 testing. After 3000 hours of 85/85 exposure, moisture in the EVA has reached a high equilibrium level across the whole cell. Figure 6.5 shows the moisture ingress for the same structure in Bangkok, Thailand. In Bangkok, the water content reaches equilibrium within ~1 year, and after that shows some seasonal variation at the cell periphery. The water concentration in Bangkok importantly reaches an equilibrium value between 0.001 and 0.0012 g/cm^3 in the center of the cell, about a factor of 5 lower than that achieved across the whole cell after 3000 hours of damp heat exposure.

In damp heat testing, the module experiences large transients in moisture content with the highest humidity levels at the cell edges (Figure 6.4). After only 250 hours of damp heat testing, the edge of the cell (~1 cm from the edge) has already reached a value in excess of 70% of its steady-state (3000 hours) concentration. On the other hand, the EVA at the center of the cell takes more than 1000 hours at 85/85 to reach the same humidity level. It is no surprise then that the EL image in Figure 5.18 shows the degradation of the module starting at the cell periphery, proceeding inward as this is how the humidity proceeds.

During field deployment in Bangkok, the moisture content at the periphery of the cell also increases more quickly than in the center (Figure 6.5). However, this is just a tempo-rary situation. After one year, the moisture content in the EVA at the center of the cell remains fairly constant, while that in the EVA at the cell periphery fluctuates seasonally about an average value comparable to the content at the center of the cell. The moisture concentration in fielded modules remains fairly constant across the surface of the cell, and never achieves the high steady-state concentration seen after long-term exposure to 85/85. If the same degradation mechanism is occurring in fielded modules that occurs in extended

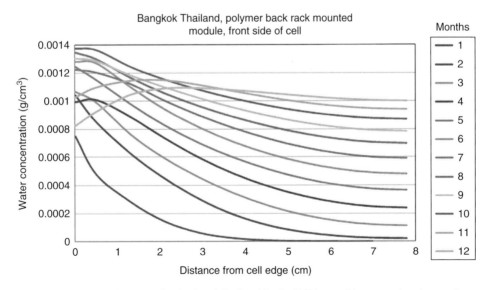

Figure 6.5 Results of analysis for the humidity level in the EVA located between the glass and a cell during the first year of exposure in Bangkok, Thailand. *Source:* printed with the permission of Michael Kempe of NREL.

85/85, we would not expect to see the same pattern of degradation in the EL signal. In the field exposed cells, the degradation of the ohmic contact would be more uniform across the cell surface.

We have looked at modules after more than 25 years of exposure and have not seen degradation of the doped oxide resulting in loss of ohmic contact. The high levels of moisture penetration shown to occur after thousands of hours of damp heat exposure (Figure 6.4) is never likely to occur in any typical terrestrial deployment. This is a failure mode that is not likely to impact module field performance.

There is a second degradation mode that is often reported after long-term humidity exposure. That is the degradation of module backsheets made from or containing polyethylene terephthalate (PET). Once again, Kempe has provided us with an analysis of PET degradation to determine whether failure of PET in long-term damp heat testing could be an issue for field exposed modules [18].

PET is commonly used as a backsheet material because of its low cost and good dielectric properties. However, as a polyester, it is susceptible to chain scission via a hydrolysis reaction. Principally, this leads to changes in mechanical properties, but loss of electrical properties can also occur due to long-term exposure to high humidity. Kempe [18] modeled the exposure of a PET backsheet and found that in one of the wettest locations on earth, Bangkok, Thailand in an insulated back-mounting system (the worst case considered) the PET backsheet is predicted to survive for 310 years. Clearly, this is far longer than the expected lifetime of any PV module. This is an important conclusion since, at one time, some module manufacturers were requiring backsheets to pass a 2000-hour 85/85 test. This turns out to be an unreasonable requirement for PET and would likely eliminate many perfectly good backsheets from use. Chapter 9 will provide more details on the PET analysis, but for this chapter we can conclude that we really shouldn't be worried about PET backsheets failing 2000 hours of damp heat testing. What we should be worried about is backsheets failing in the field due to combinational stresses that include UV exposure. This will be discussed in Chapter 7.

Finally let's address the issue of "Do it yourself" factory inspections or audit systems. There is clearly an issue with having to set up and follow a different QMS for each project. In addition, there is really no reason to think that any one of these systems is any better than the one already being used by the manufacturer. One would assume that whatever system they are using is based on their own knowledge of what steps or processes are most critical to control and have either the greatest potential to go wrong or cause the biggest issues when they do go wrong. So why would they change from their system to one recommended by the customer. Really, a manufacturer would do it only because the customer says "you must follow my rules to get my purchase order." If a manufacturer is going to make a change to their quality assurance system it should be based on assessing the system they have in place against recommendations provided by an expert group familiar with multiple PV module QMS.

The "Do it yourself" issues led to three of the major PV laboratories in the world (NREL, AIST and ESTI) working together to try to develop a better system. The story of this effort will be told in the next chapter.

References

1 ISO 9001 (2015). Quality Management Systems – Requirements.

2 Nelson, D. (2014). Misunderstanding ISO 9001. Quality Digest. https://www.qualitydigest.com/inside/quality-insider-column/misunderstanding-iso-9001.html (Accessed 29 August 2019).

3 Jordan, D.C. and Kurtz, S.R. (2013). Photovoltaic degradation rates — an analytical review. *Progress in Photovoltaics: Research and Applications* 21 (1): 12–29. https://doi.org/10.1002/pip.1182.

4 Wohlgemuth, J.H., Cunningham, D.W., Amin, D. et al. (2008). Using accelerated tests and field data to predict module reliability and lifetime. 23rd EU PVSEC in Valencia, Spain (1–5 September 2008).

5 IEC TS 62915 (2018). Photovoltaic (PV) modules – Retesting for type approval, design and safety qualification.

6 TamizhMani, M.G. (2010). Experience with Qualification and Safety Testing of Photovoltaic Modules. NREL PVMRW.

7 NREL (2012). PROPOSED TEST PROTOCOL – IEC 61215 ON STEROIDS. https://www.nrel.gov/docs/fy14osti/60169.pdf (Accessed 29 August 2019).

8 Wohlgemuth, J.H. and Kempe, M.D. (2013). Equating Damp Heat Testing with Field Failures of PV Modules. 39th IEEE PVSC in Florida, USA (16–21 June 2013).

9 Wohlgemuth, J.H., Cunningham, D.W., Amin, D. et al. (2008). Using Accelerated Tests and Field Data to Predict Module Reliability and Lifetime. 23rd EU PVSEC in Valencia, Spain (1–5 September 2008).

10 Wohlgemuth, J.H., Cunningham, D.W., Nguyen, A.M., and Miller, J. (2005). Long Term Reliability of PV Modules. 20th EU PVSEC in Barcelona, Spain (6–10 June 2005).

11 Herrmann, W. and Bogdanski, N. Outdoor Weathering of PV Modules – Effects of Various Climates and Comparison with Accelerated Laboratory Testing. 37th IEEE PVSC in Washington, USA (19–24 June 2011).

12 Saint-Lary, A., Ed-daoudi, S., Delsoi, T. et al. (2012). Photovoltaic modules reliability on accelerated and natural test. 27th EU PVSEC in Frankfurt, Germany (24–28 September 2012).

13 Peike, C., Hoffmann, S., Hulsmann, P. et al. (2013). Origin of damp-heat induced cell degradation. *Solar Energy Materials and Solar Cells* 116: 49–54.

14 Ketola, B. and Norris, A. (2011). Degradation mechanism investigation of extended damp heat aged PV modules. 26th EU PVSEC in Hamburg, Germany (5–9 September 2011).

15 Sample, T., Skoczek, A., and Field, M.T. (2009) Assessment of ageing through periodic exposure to damp heat (85°C/85% RH) of seven different thin film module types. 34th IEEE PVSC in Pennsylvania, USA (7–12 June 2009).

16 Tan, J., Ju, C., Lu, R. et al. (2017). The Performance of Double Glass Photovoltaic Modules under Composite Test Conditions. SNEC International PV Power Generating Conference in Shanghai, China.

17 Wohlgemuth, J. (2012). PV Modules: Validating Reliability, Safety and Service Life. *Proceeding of Intersolar Conference*, Munich, Germany (17–19 June 2012).

18 Kempe, M.D. and Wohlgemuth, J.H. (2013). Evaluation of Temperature and Humidity on PV Module Component Degradation. 39th IEEE PVSC in Florida, USA (16–21 June 2013).

7

The PVQAT Effort

To help address the issues with the "do it yourself" system of module quality discussed in the last chapter and to focus the PV industry on a consensus approach to improving module quality, Sarah Kurtz and John Wohlgemuth from NREL in the US, Masaaki Yamamichi and Michio Kondo of The National Institute oif Advanced Industrial Science and Technology (AIST) in Japan and Tony Sample from Joint Research Centre of European Union (JRC) in Europe with the assistance of James Amano of SEMI, organized the International PV Module Quality Assurance Forum. The Forum was held in July, 2011 in San Francisco, CA with more than 150 PV experts from around the world in attendance. The two-day Forum started with a number of presentations providing updates on what the presenters felt were the most pressing PV module reliability issues. The attendees then proceeded to discuss what was needed in order to solve those issues [1].

The Forum created a new organization entitled the International PV Quality Assurance Task Force, called PVQAT for short (Pronounced PV CAT) [2]. The forum set two goals for the new organization:

1) Development of a QA rating system that provides comparative information about the relative durability of PV modules; and
2) Creation of a guideline for factory inspections of the QA system used during PV-module manufacturing.

A third goal related to development of a conformity assessment system was added later. This third goal is:

3) Development of a comprehensive system for certification of PV systems, verifying appropriate design, installation and operation methods.

The first goal related to the fact that just using the qualification tests was not enough to ensure PV module reliability over the product lifetime as was discussed in Chapter 6. Any new testing developed under PVQAT would have to be in response to demonstrated field failures or degradation observed in module types that have successfully passed the qualification tests. So, the technical efforts started by looking at field failure modes or degradation modes related to either module wear out or to new failure modes not evaluated for in the qualification tests. PVQAT took its initial guidance from the literature, particularly from the summary work of Dirk Jordan [3]. Figure 7.1 contains a Pareto chart of observed

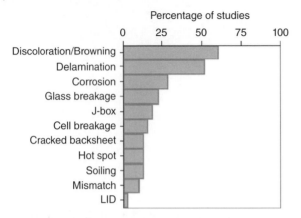

Figure 7.1 Pareto chart of module field failures [3]. *Source:* reprinted from presentation with permission of the author, Dirk Jordan of NREL.

field failure modes for PV modules. Based on these observations, PVQAT selected the following degradation/failure modes to begin researching with a goal of developing improved accelerated stress tests:

- Module discoloration
- Diode failures
- Solder bond and interconnect ribbon failures due to thermal cycling
- Potential induced degradation (PID)
- Delamination and corrosion

This analysis was the basis for the establishment of the first four module durability Project Teams (Teams 2, 3 4, and 5). Subsequent module durability project teams were added based on interest from the PV technology community.

The second PVQAT goal derives from Quality Management issues also described in Chapter 6. PVQAT efforts in this area were to focus on identifying PV specific requirements for quality management systems (QMS) to be used in the manufacturer of PV products, specifically PV modules.

The third goal derives from issues related to handling, shipping, installation and operation of PV modules in PV systems. From field reports that perfectly good modules can be damaged because of poor handling or system design, an effort was initiated to establish standards and a certification system for the PV power systems themselves. This effort will be discussed in Chapter 8.

The PVQAT website [2] says "(PVQAT) leads global efforts to craft quality and reliability standards for solar energy technologies. These standards will allow stakeholders to quickly assess a solar photovoltaic (PV) module's performance and ability to withstand local weather stresses, thereby reducing risk and adding confidence for those developing products, designing incentive programs, and determining private investments. As a result:

- Investors can gain confidence in solar investments.
- PV customers can use standards to choose products that meet their needs.
- Incentive programs can define a minimum durability for module designs.

Table 7.1 PVQAT task groups.

Group #	Task Group Name
1	PV QA Guidelines for Module Manufacturing
2	Testing for thermal and mechanical fatigue
3	Testing for humidity, temperature and voltage
4	Testing for diodes, shading and reverse bias
5	Testing for UV, temperature and humidity
6	Communications of rating information
7	Testing for snow and wind load
8	Testing for thin film modules
9	Testing for CPV
10	Testing for Connectors
11	QA for PV Systems
12	Soiling and Dust
13	Cells

- Insurance companies can adjust rates according to demonstrated reliability.
- PV-module suppliers can optimize module design to minimize cost while still maintaining confidence in reliability for a specific use or application of the modules.
- The entire PV community benefits by reducing installed PV cost when standards are created that establish durability without adding unnecessary cost."

PVQAT is open to participation by anyone who wishes to contribute. The efforts rely on research done by volunteers around the world. PVQAT helps guide worldwide research to answer important questions related to testing of PV products that predict outdoor performance of PV modules. The objectives of this work are to develop International Electrotechnical Commission (IEC) standards that can be used for testing PV modules that can then be deployed in any terrestrial environment. This is to replace the "do it yourself" quality system. To implement this work, PVQAT has established a number of working groups. Table 7.1 contains a list of the 13 working groups established to date. Each of these groups has been active during at least part of the time since PVQAT's formation in 2011. Subsequent subsections will discuss the work of each of these task groups.

7.1 Task Group 1: PV QA Guidelines for Module Manufacturing

Task Group (TG) 1 actually consisted of four groups, one each in the US, Europe, Japan and China. Each of the groups began their effort by reviewing the ISO 9001 standard. Each made recommendations about how to improve the generic ISO 9001 for PV-module manufacturing. Once each of the groups had finished reviewing the document, a worldwide team under the direction of Paul Norman from the US and Ivan Sinicco from

Switzerland began to review and combine the four lists into one coherent document. These were incorporated into a document entitled "Proposal for a Guide to Quality Management Systems for PV Manufacturing: Supplemental Requirements to ISO 89001–2008" which was published as a National Renewable Energy Laboratory (NREL) report [4].

In reviewing the generic requirements in ISO 9001, TG 1 considered the following items important in its translation to the manufacture of PV modules:

- A PV-module manufacturer should document its control of the PV module's design and show that the design is aligned with the expected lifetime and with the product warranty.
- Handling of warranty claims must be addressed by product and process design and/or by financial means.
- Potential failure modes should be considered and steps taken to address them (e.g. through a Failure Modes and Effects Analysis, FMEA) in design, production and the delivery process.
- Products should be IEC certified to Qualification and Safety.
- All PV-module manufacturers should have an ongoing reliability test program that monitors PV-module performance for compliance with standards and the stated design lifetime.
- Product traceability should be implemented throughout the entire supply chain to enact positive control of the product for recalls and warranty claims.
- The manufacturing process should be designed to ensure conformance to the design intent for performance, lifetime, and warranty
- Some special processes such soldering and electro-static discharge (ESD) control, should be detailed within the quality program.
- PV-module power ratings must be made in conformance with IEC standards and the quality system must document the appropriate controls.

Based on this work, TG 1 published a paper describing the requirements of the guidelines for PV-module manufacturing [5].

The next step was to turn this into an IEC International Standard (IS) or Technical Specification (TS). This is where IEC politics got in the way. IEC has two main branches, the Standards Development Organization of which TC82 is one of the Technical Committees and the Conformity Assessment Organization, where systems are established to determine whether a product, system or service corresponds to the requirements contained in a specification. An IEC Technical Committee like TC82 for PV cannot write a conformity assessment standard. So, the Conformity Assessment part of IEC said that TC82 could not write a standard called "ISO 9001 for PV-module manufacturing." However, a Technical Committee has every right to write a standard to assist manufacturers in meeting the requirements of one of its own standards. So TC82 proposed a new work item for a TS entitled "Guideline for increased confidence in PV module design qualification and type approval." This was accepted by the IEC Central Office and approved for development by the National Committees within TC82.

A combined team that included PVQAT TG 1 led by Rovind Ramu from the US and the TC82 Working Group 2 Project Team led by Yoshihito Eguchi from Japan worked diligently to turn the NREL report into a TS for IEC. The group set a speed record for IEC as their document was approved as IEC TS 62941 in January, 2016 [6]. This TS is now available for manufacturers to use in certifying their QMS. A number of PV-module manufacturers have already completed the process.

PVQAT TG 1 did not consider its work complete with the publication of IEC TS 62941. As IEC TS 62941 moved toward publication they updated the NREL Technical Report to keep the two consistent [7]. Then they began working on developing aids for the audits that are required for certification to IEC TS 62941. They are also working on setting up a training program for auditors and are preparing a checklist to be used by the auditors to help guide them through the audits.

Because IEC TSs are only valid for three years, Working Group 2 of IEC TC82 has already begun the work to update IEC TS 62941 and to turn it into an IS. Along with the change to an IS, the Project Team is proposing a name change to "Terrestrial photovoltaic (PV) modules – Quality system for PV-module manufacturing." A Committee Draft of this new document has already been prepared and reviewed by the National Committees within TC82.

IEC TS 62941 is an important tool for use in the IEC System for Certification to Standards Relating to Equipment for Use in Renewable Energy Applications (IECRE) Conformity Assessment System that will be discussed in Chapter 8.

7.2   Task Group 2: Testing for Thermal and Mechanical Fatigue

As already stated in Chapter 2, failures of cell interconnect ribbons and solder bonds are important failure modes for PV modules. The failure rates are dependent on thermal and mechanical stresses. TG 2 is studying how to induce and quantify these failures. The Module Qualification Test IEC 61215 already contains 200 thermal cycles (from $-40\,°C$ to $+85\,°C$ with maximum-power current flow when the modules are above room temperature) to detect such failures. Field data indicate that module types that survive this 200-thermal cycle test have still suffered broken interconnects and degraded solder bonds in the field over times that are significantly less than the warranty period [8]. Therefore, the 200-thermal cycle test may not provide enough stress to validate modules for a 25-year warranty in all environments. The main questions then are how many thermal cycles should be used for the test and should the cycles be the same for the different climate zones and mounting configurations?

Extended testing indicates that modules susceptible to thermal-cycle failures begin to show significant power loss between 400 and 600 cycles [9, 10] Figure 7.2 shows the results from extended testing [9] indicating that not all commercially available modules qualified to IEC 61215 can survive 400 thermal cycles without power loss. A number of PV-module manufacturers including BP Solar [11] have reported on using 500 thermal cycles for internal quality control so this value has been put forward as a starting point for an extended thermal-cycle test.

There is also the question of whether the temperature range and rate of temperature change should be modified? There are certainly locations (think, for example, roof tops in Arizona) where the maximum module temperature will often exceed 85 °C. Would it help to raise the upper temperature range for such locations? Can the rate of temperature change

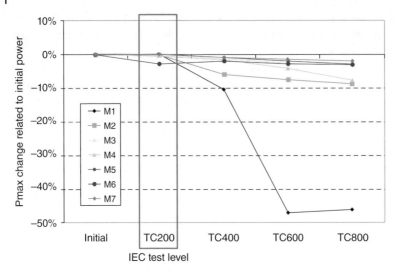

Figure 7.2 Power loss as a function of thermal cycles [9]. *Source:* reprinted from PVSC paper with the permission of the article's author, Werner Herrmann of TÜV.

be increased to both increase the stress as well as to accelerate the number of cycles that can be performed in a given time?

The Leaders of TG 2 (Nick Bosco from NREL, Tadanori Tanahashi from AIST and Simon Xiao from Trina Solar) took on the challenge to model the weather-induced stresses on solder bonds and interconnect ribbons to better understand what parameters are responsible for the stresses induced in the modules.

Bosco's first effort involved trying to use dynamic mechanical loading to break interconnect ribbons in much shorter time periods than it takes using thermal cycling. He was successful at causing interconnect breakage using the cyclic (dynamic) loading technique from IEC TS 62782 [12] using much higher pressures than the 1000 PA specified in the standard [13]. However, he was not able to correlate the test results from the dynamic mechanical loading because he was not able to get any of the thermal cycle samples to fail after several thousand cycles even though he had samples specifically built to fail prematurely with interconnect ribbons soldered to the edges on both front and back of the cells. His conclusion was that the newer, lower stress interconnect ribbons were unlikely to break in the field.

Therefore, he shifted his efforts to evaluating solder bond failures. He was guided in this work by several reports from the field indicating that fill factor (FF) degradation of modules was dependent on the climate where they were deployed. During an all-India field survey [14], it was observed that modules deployed in hot climates exhibited power degradation due to an increase in series resistance, while modules deployed in cooler climates had less power degradation and exhibited little or no increase in series resistance. Jordan made a similar observation after analyzing the data from more than 2000 PV systems from around the world [15]. One of the major causes of increased series resistance is failed solder bonds so this is certainly one possible explanation for why modules tend to suffer higher degradation rates in hotter climates.

Table 7.2 The seven cities used in Bosco study [16].

City	Climate	Mean monthly maximum Temp (C)
Phoenix, AZ	Hot and dry	38
Chennai, India	Hot and humid	37
Tucson, AZ	Hot and dry	36
Bhogat, India	Hot and humid	35
Honolulu, HI	Hot and humid	31
Golden, CO	Temperate	27
Sioux Falls, SD	Cold	23

Bosco then attempted to model the stress levels on solder bonds as a function of climate to see if his model predicts more solder bond degradation in hotter climates. He began his effort by developing a model (FEM) of a flat-plate terrestrial module in order to calculate the accumulation of inelastic strain energy density or damage within the solder joints from either field exposure or thermal cycle testing [16]. He created a two-dimensional model of a standard cry-Si PV module. The construction consisted of:

- A glass superstrate 3.6 mm thick
- Encapsulated in 450 μm of EVA
- With a 175 μm thick single layer generic polymer backsheet
- Silicon solar cells 175 μm thick
- Interconnect ribbons 150 μm thick
- PbSn solder bonds 30 μm thick

Properties of these materials were taken from the literature.

Simulations were run for thermal-cycling tests and for the weather in seven selected cities spanning climates from hot to temperate to cold. The seven cities are listed in Table 7.2 along with their climate rating and their mean monthly maximum temperature. The King model was used to calculate the module (actually cell) temperature from the ambient temperature, irradiance and wind speed [17]. In this way a one-year module temperature profile was created using the meteorological data from each of the seven cities.

Figure 7.3 shows the calculated value of damage that accumulates in the solder bonds over the course of one year in each of the seven cities. Chennai, India accumulates the most damage although it is not the hottest city. The other two cities with hot and humid weather, Honolulu and Bhogat, have considerably less damage. Indeed Golden, Colorado, a temperate location, has more accumulated damage than either of these two hot and humid environments. Tucson and Phoenix have similar curves but Tucson has more accumulated damage although its mean monthly maximum temperature is less than that of Phoenix. We can tell from this figure that mean maximum temperature is not the only variable impacting the amount of damage.

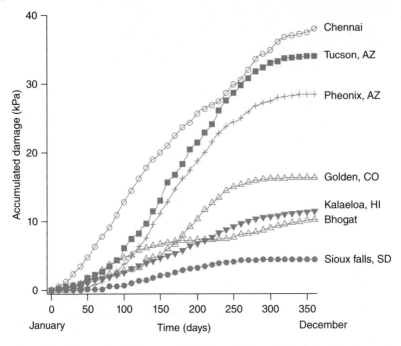

Figure 7.3 Accumulated damage within the solder bonds in one year in the seven selected cities [16]. *Source:* reprinted from **Microelectronics Reliability** under the STM Permission Guidelines and with permission of the author of the paper, Nick Bosco of NREL.

Figure 7.3 is plotted in such a way that the slope of the curve represents the damage rate. A steeper curve means more damage is occurring over that time period. The maximum rates (steepest curves) are all similar for Chennai, Tucson, Phoenix and even Golden. This means that during the hottest times of the year the damage rates are similar. Chennai just stays hotter longer while Golden has the shortest summer. In cold Sioux Falls, almost all of the damage occurs during the summer with the curve being flat during the colder months. The FEM analysis shows that it is the softening of the module's encapsulant at high temperatures that results in a faster rate of solder damage accumulation.

The FEM analysis is interesting, but it really doesn't tell us what the controlling factors in the weather are nor is it going to be convenient to run the model for every city we are interested in. Therefore, Bosco set out to develop an empirical model to calculate the damage in a way that would agree with the FEM model. His empirical or climate model is based on the well-established Coffin-Manson and Norris-Landzberg equations [18]. The model for calculating the damage, D, from the weather data takes the form given in Eq. (7.1)

$$D = C(\Delta T)^n (r(T))^b \exp\left\{-\frac{Q}{k_B T_{max}}\right\} \tag{7.1}$$

Where:

ΔT = the mean daily maximum module temperature change
T_{max} = the mean daily maximum module temperature

C = a scaling constant

Q = an Activation Energy

k_B = Boltzmann's constant

r(T) = a temperature reversal term, which is equal to the number of times the temperature history increases or decreases across the reversal temperature, T, during the course of one year.

The values of the exponentials *n* and *b* as well as the activation Q are the values for PbSn eutectic solder taken from the Norris paper [18]. Bosco then used the scaling constant C and the reversal temperature T to fit the climate model to the FMEA model. The results are shown in Figure 7.4. The fit is good for weather data intervals of 30 minutes or less. All attempts to fit without a temperature reversal term ended in poor agreement.

This result shows how important the reversal term is. This makes sense as the reversal term represents heating and cooling of the module during the day due to variability of the irradiance. Locations with more intermittent clouds will have more reversals and therefore more thermal-cycling stress on the solder bonds. What is interesting from this result is the fact that the reversal temperature that yields the best fit is 56.4 °C. If the module never reaches 56.4 °C during the day any temperature reversals will have minimal effect. That is why Sioux Falls had very low damage and why Golden only has damage during the summer months. So, the climates that see the most damage to their solder bonds are typically characterized by high ambient temperatures and partly cloudy days.

Table 7.3 shows the important weather characteristics for the seven cities studied. You can see how the amount of damage to the solder bonds is associated with the number of reversals across 56.4 °C occurring during the year. This data also shows that Tucson has more temperature reversals than Phoenix so modules suffer more damage to solder bonds

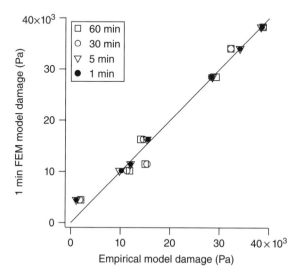

Figure 7.4 Comparison of damage to solder bonds calculated using finite element modelling (FEM) simulation and modeled by Eq. 7.1 for 4 weather data intervals [16]. *Source:* reprinted from **Microelectronics Reliability** under the STM Permission Guidelines and with permission of the author of the paper, Nick Bosco of NREL.

Table 7.3 Weather characteristics for seven cities used as input to empirical model [16].

City	T_{max} Mean Daily Max Temp (C)	ΔT Mean daily max Temp Change (C)	r(56.4 °C) number of temp reversals across 56.4 °C
Chennai, India,	59.0	35.5	2060
Phoenix, AZ	53.4	34.7	1298
Tucson, AZ	53.2	37	1592
Honolulu, HI	51.2	30.7	212
Bhogat, India	49.7	27.9	234
Golden, CO	37.6	33.4	518
Sioux Falls, SD	25.8	23.5	2

in Tucson than Phoenix even though Phoenix is hotter. Similarly, Golden is cooler than Honolulu and Bhogat, but a module in Golden will experience more temperature reversals than it would in either of the other two cities and, therefore, it will suffer more damage to its solder bonds.

The final thing to do is to compare the damage to solder bonds calculated using the weather in the seven cities with the damage that occurs during thermal-cycling testing. Bosco looked at both the standard thermal cycles (−40 °C to +85 °C) and what happens if you increase the range of temperature cycling. He saw little to no impact in the accumulated damage in the solder bonds by changing the lower temperature limit. However, the damage level could be changed significantly by changing the upper temperature limit. Figure 7.5 shows the number of thermal cycles required to cause equivalent damage to a one-year exposure in the seven cities.

There is a huge difference between cities. For Chennai, it takes 25 standard thermal cycles (−40 °C to +85 °C) to create the same damage as one year of field exposure. On the other hand, for Sioux Falls it only takes three to four standard thermal cycles to create the same amount of damage as one year in the field. So, the 200 thermal-cycle test in the Qualification Test sequence (IEC 61215) should be adequate for a 25-year lifetime for an open-rack mounted system deployed in Sioux Falls, South Dakota or any other cooler place, probably including all of northern Europe. On the other hand, 200 thermal cycles is not enough testing for Chennai, Tucson or Phoenix. Testing for a 25-year lifetime for these locations would require at least 500–600 standard cycles (−40 °C to +85 °C).

Figure 7.5 also shows that there is an option to use a higher maximum temperature with fewer cycles to achieve the same stress level. For Chennai five cycles from −40 °C to +100 °C produce equivalent stress as 1 year in the field. So, 125 of these cycles may be enough to predict 25-year survival against solder bond failures in Chennai and probably in any other terrestrial location. As far as we know, there hasn't been much research done to evaluate the effects of increasing the upper limit of cycling temperature. This is something that needs to be done so that fewer, more severe cycles can be used to more quickly qualify modules for solder-bond survival in the harshest terrestrial environments.

In a subsequent paper, Bosco evaluated the impact of materials and their design on the results of his damage modeling [19]. He found that variables like solder thickness

Figure 7.5 Number of thermal cycles required to cause damage equivalent to one year in the field [16]. *Source:* reprinted from **Microelectronics Reliability** under the STM Permission Guidelines and with permission of the author of the paper, Nick Bosco of NREL.

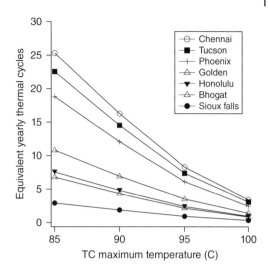

(above a certain minimum), copper-ribbon thickness and silicon thickness had little effect on the results of his modeling. Of course, use of non-solder bonds would make a big difference as would use of an encapsulant with elastic properties very different than ethylene vinyl acetate (EVA). However, even for these materials, the general findings of high temperature reversals being the main driving force behind the damage and that more thermal cycles (than in IEC 61215) should be performed for these environments will likely hold.

The last step in this process is to turn the TG 2 work into an IEC standard. This process is now complete as IEC 62892 [20] was published in 2019. The test in this proposed standard is based on the ability for 95% of the modules represented by the samples submitted for the test to pass an equivalency of 500 thermal cycles of −40 °C to +85 °C. The baseline is to test two modules through 731 cycles of −40 °C to +85 °C. However, the number of cycles can be reduced by either increasing the upper limit in the temperature cycle or by increasing the number of modules tested. There is a table in the standard that defines how many cycles have to be performed based on the selected maximum cycle temperature and the number of modules tested. As an example, if you stay with two modules but increase the maximum cycle temperature to 95 °C, the number of required cycles reduces to 563. Increasing the number of modules tested from 2 to 8 (without changing the −40 °C to +85 °C range) reduces the required number of cycles from 731 to 580. So, the proposed standard provides options for shortening the test time.

7.3 Task Group 3: Testing for Humidity, Temperature and Voltage

TG 3 is focused on testing for humidity, temperature, and voltage stresses. The ingress of moisture with or without electrical bias has been shown to cause corrosion and charge/ion transport in PV modules. Temperature and humidity have been used as accelerated stress tests for PV modules for many years. There are multiple humidity and humidity/electrical bias degradation modes with widely varying acceleration factors. The leaders of TG 3 (John Wohlgemuth of NREL, Neelkanth Dhere of FSEC, Peter Hacke of NREL, Tadanori Tanahashi of AIST and Alan Xu of Canadian Solar) started by assessing the performance of field-aged modules. They collected and continue to collect performance data from modules with more than 10 years of exposure in hot humid climates in an effort to identify humidity-related failure and degradation modes.

The present Qualification Test (IEC 61215) already includes both a damp heat test (1000 hours at 85% RH and 85 °C) and a Humidity-Freeze Test. There is no question that when use of these tests was implemented, the resultant modules had much improved durability to degradation due to humidity and temperature. When the damp heat test was first introduced into the Qualification Test sequence, it was the hardest leg to pass. Often one or more of the test modules suffered more than 5% power loss and required retesting. However, over the years, the module manufacturers made improvements that led to their modules more easily passing the damp heat test. Some of the things that contributed to this improvement include more consistent EVA with primer incorporated, improved screen-print pastes, better backsheets, and improved control over the processes particularly lamination to ensure both adequate cross-linking and adhesion of the encapsulant to the other surfaces within the module. For at least the last 10 years, most cry-Si module types actually gain one percent or two in power after the damp heat test.

TG 3 sought to answer the question "what failure modes are being seen that could be related to humidity but are not caught by the damp heat or humidity-freeze tests and what do those failure modes look like?" From the list of overall failure modes given in Chapter 2, there are only a few that are attributed exclusively to high humidity levels. Let's take a look at each of these.

7.3.1 Corrosion

Corrosion of cells, cell metallization, interconnect ribbons, bus bars and connectors can all be facilitated by moisture. As stated previously, it is seldom the semiconductor itself that corrodes but rather the transparent conductive oxides (TCO) or metallization. The solution to having moisture-sensitive material within the package has been to provide a sealed package with robust moisture barriers, including glass substrates and superstrates as well as an edge seal, to keep the moisture from penetrating into the module. Such packages are designed to easily withstand several thousand hours of damp heat and hopefully 25 years in the field to fulfill the manufacturer's warranty. For this package geometry, the critical factor in determining long-term survival in the field is whether the edge seal can survive the stresses of the environment (UV, thermal stresses, mechanical stresses, etc.).

There were cases of infant mortality of modules resulting from corrosion of the cell metallization observed in the early days of PV technology. However, after introduction of the damp heat and humidity-freeze test into the Qualification test procedure, the number of observed field failures due to corrosion decreased significantly. This suggests that some of the major issues related to corrosion field failures have already been solved by the humidity-freeze and damp heat tests now incorporated into the qualification protocol.

7.3.2 Delamination

Delamination of the encapsulant has been identified as one of the most prevalent field-failure modes for PV modules in several studies [3, 21] as was shown in Figure 7.1. When delamination is accompanied by corrosion of the cell metallization or interconnect ribbons, it can lead to significant power loss. Encapsulant delamination has been observed from both the front glass and from the cell/metallization surfaces.

1) <u>Delamination from glass</u>: Encapsulant-glass delamination has been observed in specific module types. NREL has reported on such delamination as shown in Figure 7.6 in Mobil Solar and later ASE America modules at a number of sites with a variety of different climates [22]. These modules were fabricated with a non-EVA encapsulant which did not have as high an initial adhesion to glass as some other encapsulants like EVA do. This material tended to delaminate from the front glass either directly over the junction box or in the corners of the module. This problem was observed at numerous sites in both humid and dry locations indicating it is probably not primarily driven by humidity.

   Most EVA-encapsulated modules do not exhibit this type of Encapsulant-Glass delamination. To help us understand why, Nick Bosco of NREL utilized a recently developed cantilever beam method to measure the adhesion between EVA and glass in old Arco Solar modules – comparing the results obtained from modules exposed in the field for 27 years with those obtained from a module continually stored in a shed for the same amount of time [23]. For the sample that had been in a shed for the whole time, the weakest interface was the EVA-cell interface with a debond energy (adhesion) of between 800 and 1000 J/m^2. For the sample that was exposed in the field to sunlight for 27 years, the weakest interface was the glass-EVA interface with an adhesion of between 200 and 300 J/m^2. When measuring the adhesion of EVA to glass in new samples we

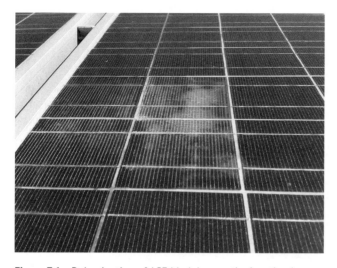

Figure 7.6 Delamination of ASE Module over the junction box.

typically obtain values of around 2000 J/m^2. So, the front-surface adhesion has dropped by about an order of magnitude. These modules show no evidence of delamination and we would expect that a debond energy of 200 and 300 J/m^2 between encapsulant and glass is adequate for module survival. Some recent work to be discussed in Section 7.5 on UV testing will provide additional information as to why this failure mode should not be a major issue for PV going forward.

2) <u>Delamination from cells:</u> For EVA-based modules, the more common observation is delamination between the EVA and the cell surface selectively occurring around the interconnect ribbons as shown in Figures 2.6 and 7.7 for Siemens Solar modules. This failure mode tends to occur in a large fraction of the modules in certain arrays. The modules in Figures 2.6 and 7.7 were deployed in Florida [22], where all of the modules in the array exhibited a similar level of delamination. However, there are numerous Siemens Solar arrays with little or no delamination. It is interesting to note that such delamination appears in multiple installations from Arco Solar, Siemens Solar and even Shell Solar indicating that the source of the delamination was due to the module design or manufacturing process for many years. Because this delamination involves the grid lines, it is almost always accompanied by loss of power due to a reduced fill factor.

As far as we know, delamination between cells and encapsulant as shown in Figures 2.6 and 7.7 has never been reported to occur due to the accelerated stress tests in IEC 61215. Such a delamination has not been seen after thermal cycling, humidity freeze, damp heat or even extended damp heat. So, it does not appear to be the result of a single stress. TG 3 began to explore possible combined-stress sequences that might lead to such delamination occurring. The first attempt was to build on the UV/50TC/10HF test leg in IEC 61215 by increasing the UV portion to a level that would significantly degrade the adhesion between the EVA and cells [24]. The amount of UV exposure required to reach a plateau in adhesion was provided by the work of TG 5 on UV, Temperature and Humidity (see Section 7.5) [25].

The whole test sequence consisted of:

- UV exposure to 0.4 MJ/m^2 at 80 °C.
- Cyclic (Dynamic) Mechanical Loading Test per IEC TS 62782.
- 50 Thermal Cycles from −40 °C to +85 °C per IEC 61215-2 MQT 11.
- 10 Humidity-Freeze Cycles per IEC 61215-2 MQT 12.
- 1000 hours of damp heat at 85%RH and 85 °C per IEC 61215-2 MQT 13.

To validate any such new test sequence, one has to identify a module technology that would pass IEC 61215 but fail the new sequence and then would fail in the field. To test

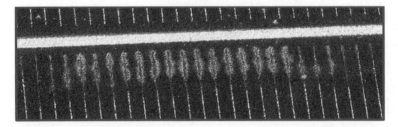

Figure 7.7 Delamination in siemens solar module. *Source:* picture provided by David Miller of NREL.

this concept, TG3 needed to develop a method of building modules that can pass the qualification test but fail this sequence. In an attempt to do this, NREL built modules with three different kinds of EVA consisting of the 15295P/UF formulation (a traditional fast-cure formulation from STR) with (i) the standard amount of Dow Corning Z6030 primer, (ii) 20% of the standard amount of primer and (iii) no primer at all. These were tested through IEC 61215 and the delamination test sequence described above. As expected, the modules built using EVA without primer failed IEC 61215 (failing both damp heat and the UV/TC/HF leg) and failed the delamination test sequence, delaminating during the initial extended UV exposure. As expected, the other two EVAs (standard and 20% primer levels) passed IEC 61215. Also as expected, the 100% primer mini-modules successfully passed the new delamination test sequence. However, the mini-modules with 20% primer also successfully passed the extended delamination test sequence with no observable delaminations. Even reducing the level of primer below 20%, but still adding some (~ 10% of standard) resulted in mini-modules that passed the delamination cycle. It seems like adding more UV did not facilitate delamination like we thought it might. This is likely due to the fact that the EVA type in these mini-modules (as well as in the Arco Solar modules) is filled with UV absorbers, so not much if any at all of the UV probably reaches the EVA-Si interface.

So, trying to combine a set of tests with humidity, temperature, mechanical stress and UV has not been successful at reproducing the EVA-cell delamination seen in the field. The one thing left out of that cycle is an electric field and/or current flow, something always associated with PV modules working in the field. So, let's go ahead and discuss the one failure mode that we know involves voltage, that is PID and come back and discuss delamination again at the end of that subsection.

7.3.3 PID

As indicated in Chapter 2, PID is a newer failure mode that is not addressed in the present qualification test sequence. At the International PV Module Quality Assurance Forum in 2011, PID was listed as one of the areas of concern so TG 3's charter was developed to include humidity, temperature and voltage. Soon after the Forum, Working group 2 of IEC TC 82 created a Project team under Peter Hacke of NREL to develop a standalone TS to define a test method or methods for evaluating a module's susceptibility to PID degradation. The standards community felt it was better to first publish a TS with just one or several test methods for the industry to evaluate rather than to try to immediately incorporate a PID test into the qualification test sequence.

The efforts to develop such a test resulted in two quite different test methods evolving. After much discussion at the international level, it was decided to include both methods in the IEC TS, IEC TS 62804-1 [26]. The two methods are:

1) Contacting the surfaces with a conductive electrode: In this method, a conductive foil (like Al foil) is placed over the front-glass surface and the frame of the module if it has one. The entire test is performed under laboratory conditions (25 °C, less than 60% relative humidity). The rated module system voltage is then applied between the grounded foil and the cell circuit. The recommended duration of the test is 168 hours. This test is supposed to simulate a condition where the front glass is coated with water. It will defeat

any designs that attempt to limit PID via frame or mounting system design. It is a good quick test to determine whether a module is susceptible to PID.

2) <u>Environmental Chamber method</u>: In this method, the module is placed in an environmental test chamber and exposed to 60 °C and 85% relative humidity. Once the module has stabilized under those conditions the rated module system voltage is applied. The recommended duration of the test is 96 hours. Once again, this is a quick test that determines whether a module is susceptible to PID. In this case, however, methods using framing and mounting to reduce PID can be evaluated. The TS also allows for increasing the chamber temperature to 65 °C or even 85 °C if desired.

As the industry has had time to use these two approaches, it has become apparent that they do not yield the same results though they are both good at identifying modules susceptible to PID. It is likely that they are assessing what happens under different conditions in the field. The Al foil test is probably a good simulation of what happens when a module's surface is wet. The chamber test is probably a good simulation of hot, humid days in the field. There have been a number of publications trying to compare the accelerated stress test with outdoor performance in various climates. Hacke [27] compared the results of the 60 °C and 85% relative humidity chamber test with outdoor performance in Florida. His results indicated that the modules that failed the chamber test also failed quickly in the field while those that passed the chamber test survived at least through the duration of the outdoor exposure.

Hoffman [28] demonstrated that a plot of module leakage current at 1000 V bias versus inverse temperature resulted in a straight line regardless of whether the module was exposed outdoors, in a humidity chamber or with an Al foil over the glass. In all cases, the straight lines were parallel; indicating the same activation energy meaning the same phenomena was responsible for the behavior. The leakage current at 60 °C and 85% RH was three times that of the Al foil measurements at 25 °C [29]. Using this concept any outdoor exposure could then be modeled using data from indoor tests (chamber or foil) at different temperatures to represent the different time periods during the year.

The cry-Si PID TS was published in 2015. Industry has now been using the different PID tests for a number of years. Over time, many of the PV-module manufacturers have elected to go with the chamber test at 85 °C/85% RH for 96 hours. This is much more stressful than the 60 °C/85% RH in the TS. So, if a manufacturer builds modules that pass the higher stress level chamber test, they should easily pass the lower stress level chamber test or the foil test. WG2 is planning to add the PID test for cry-Si modules to the next edition of IEC 61215. The higher level of stress (85 °C/85% RH) chamber test has been proposed (see Chapter 10.) Only time will tell if it stays that way through the IEC approval process.

Once IEC TS 62804-1 for cry-Si was published, efforts turned to developing IEC TS 62804-2 for thin-film modules. Recent philosophy within the standards community has been that if at all possible, test methods should be the same for all technologies. However, in this case many thin-film module types use double-glass construction with edge seals so are resistant to moisture ingress. A 96-hour humidity exposure is not likely to cause any moisture ingress or degradation in such a case. The question then becomes how much longer should a thin-film PID test be? While there has been much discussion this is still an open question.

7.3.4 Delamination Due to Voltage Stress

As stated earlier, the one thing that had not been included in the proposed PVQAT combined stress sequence for delamination was applied voltage. The earlier work was all done with no applied voltages. As part of the NREL research effort under Peter Hacke to better understand PID, mini-modules that had previously been subjected to 1000 hours of damp heat at 85 °C/85% RH, were left to dry for an extended period of time before being subjected to a PID test at 72 °C/95% RH with a negative bias of 1000 V for 196 hours. As can be seen in Figure 7.8, areas of delamination appeared, similar to those seen in the Siemen Solar modules from the field shown in Figures 2.7 and 7.7 [24].

This experiment has been repeated on several mini-modules using various times for the PID exposure after the initial 1000 hours of damp heat. After a PID stress time of 110 hours, only a very small amount of delamination was visible. After an additional 137 hours of continued PID stress, the delaminated area significantly increased. The next step was to repeat the test protocol on commercially available cry-Si modules. Hacke [30] selected four types of commercial modules with varying PID behavior. He then subjected them to 1000 hours of damp heat (85 °C/85 %RH) followed by a PID test. The results are summarized in Table 7.4.

From Table 7.4 we see that while the three module types with the highest leakage current all suffered from PID power loss, only module type 3 with the highest leakage current

Figure 7.8 Delamination between encapsulant and cell after DH/PID stress testing [24]. *Source:* reprinted from author's PVSC paper.

Table 7.4 Results of combined DH/PID stress test on four types of commercial modules.

Module Type	PID Power Loss	Delamination	Leakage Current At −1000 V (Amp)
1	Yes	No	$6.4 \cdot 10^{-5}$
2	Yes	No	$2.6 \cdot 10^{-5}$
3	Yes	Yes	$4.4 \cdot 10^{-4}$
4	No	No	$7.9 \cdot 10^{-7}$

suffered delamination. While it is too early to identify the exact cause of the delamination, preliminary thoughts are that an electrochemical reaction occurs at the metallization-encapsulant interface resulting in generation of a gas (maybe oxygen and hydrogen from the water in the encapsulant) and that this gas expands pushing the encapsulant away from the Si and metal surface.

While these are preliminary results, it is encouraging that we can finally reproduce the type of delamination seen in the field. Based on this work, a new IEC TS (IEC TS 62804-1-1) for PID delamination is under development.

7.4 Task Group 4: Testing for Diodes, Shading and Reverse Bias

TG 4 is focused on testing bypass diodes, which are essential to protecting the module during partially shaded conditions. If bypass diodes fail in open circuit, partial shading can cause severe hotspots and fire hazards. For commercial 60 or 72 cell cry-Si modules, failure of one bypass diode in short-circuit results in approximately one-third reduction of output power, certainly a module failure. Even with working bypass diodes, continuous operation of cells in reverse bias means the diodes will operate at high temperatures over long time periods which can cause permanent degradation of the diodes.

The details of the stresses that lead to these diode failures are not very well understood. The leaders of TG 4 (Vivek Gade of Jabil, Narendra Shiradkar formerly of Jabil and now with the Indian Institute of Technology Bombay, Paul Robusto of Miasole, Yasunori Uchida of JET, Xian Dong of Zhongshan University and Chandler Zhang of Hohai University) are working toward developing enhanced understanding of the stressors and field-failure mechanisms in commonly used silicon-based Schottky bypass diodes. Knowledge of long-term, shading-induced degradation mechanisms in cells and bypass diodes in the field is essential for estimating the service life of PV modules.

TG 4 has worked on four specific projects that will be discussed below.

1) <u>Diode functionality in Qualification and Safety Tests</u> – When modules are put through the accelerated stress tests in IEC 61215 and IEC 61730 the bypass didoes in the module and/or the junction box will also be subjected to the stresses. If a bypass diode was to fail via a short circuit it would cause a significant power loss and therefore be flagged as a

failure of the test. However, if a bypass diode were to fail open circuit it would not affect the output power of the module. With an open-circuited bypass diode, the module will work fine but be susceptible to reverse bias hot spot damage. In the earlier versions of the Qualification and Safety documents (IEC 61215: 1993 and 2005, IEC 6646: 1996 and 2007 and IEC 61730-1 and 2: 2004), there was no requirement to validate that the bypass diodes had not failed open circuit due to the imposed stresses. To remedy this, a Bypass Diode Functionality test was developed and included as MQT 18.2 in the 2016 version of IEC 61215. However, this standard only requires the Bypass Diode Functionality Test to be performed after Sequence B (see Figure 4.3) which includes the Bypass Diode Thermal Test and the Reverse Bias Hot Spot Test. This was remedied in the second edition of IEC 61730 which refers to the test in MQT 18.2 and requires it to be performed on all of the tested modules as part of Group B of the final measurements (see Figure 4.7). The plan moving forward is to include a requirement to validate bypass diode functionality at the end of all test sequences in the next edition of IEC 61215 (see Chapter 10).

2) <u>Electrostatic Discharge (ESD) Failures</u>: It is well known that diodes, especially the Schottky diodes used as bypass diodes in PV, are susceptible to damage by ESD and that ESD damage to bypass diodes can lead to open-circuited diodes and premature module field failures [31]. Kent Whitfield led a WG2 team to develop IEC TS 62916 [32] which provides a means to evaluate the susceptibility of bypass diodes to fail due to ESD. ESD may occur whenever there is contact or close proximity between the diode and an object of different electrostatic charge. Of interest in this TS are relatively low energy, short-duration surges that may occur during the manufacturing process, during testing, or during installation when bypass diodes may directly exposed to ESD stresses. The TS provides a test method where the bypass diodes are subjected to progressive ESD stresses. The document then provides a means for analyzing and extrapolating the resulting failures to determine diode surge voltage tolerance to ESD. This can then be used by module manufacturers to set limits on the allowable ESD in their factory.

3) <u>Bypass Diode Thermal Runaway</u>: In normal operation, the bypass diodes within a PV module are reverse biased and not passing current. When the module is shaded and some of the cells cannot produce the current being produced by the other cells in the series string, the shaded cell or cells are driven into reverse bias and the bypass diode protecting them turns on. Under these circumstances, the bypass diode is passing significant current and therefore its temperature increases significantly. It is this condition that is tested in MQT 18 of IEC 61215-2, the Bypass Diode Thermal Test. When the shade is removed, the operating conditions of the diode will return to normal and it is again reversed biased.

For many diode types, the reverse leakage current increases with temperature. When conducting in a forward direction the diode may be at a fairly high temperature due to heating by the sun and by the current flow through it. If the shadow goes away, the diode will switch to being reverse biased. If the leakage current of this heated diode is too high and the thermal design of the junction box is not adequate, it will continue to heat up and go into "thermal runaway." In "thermal runaway" the diode just continues to heat until it destroys itself.

It is the thermal design of the bypass diode in the junction box that determines whether it will survive or suffer thermal runaway. IEC 62979 [33] was written to provide a test method for determining whether a particular diode/junction box combination is safe or if it has the potential for thermal runaway. In this test, the diodes are heated in a chamber to 75 °C for open rack mounted or to 90 °C for other applications like roof mounted for 40 minutes (or longer if necessary, to stabilize the temperature) with a current equal to 1.25 times the short-circuit current of the module running through them. Then the current is shut off and within 10 milliseconds a reverse bias voltage equal to the open-circuit voltage of the string of cells protected by the diode is applied. The temperature of the diode is monitored. If the temperature of the diode decreases and it still functions as a diode, it passes the test and is considered safe for use in a PV module. If the temperature of the diode increases or if it no longer functions as a diode afterward, it fails the test.

4) <u>Higher Temperature Operation:</u> The Bypass Diode Thermal test (MQT 18.1 in IEC 61215-2) specifies a test at 75 °C. In part 1 of the test, the diode is operated at the STC Isc of the module at 75 °C for one hour. To pass, the measure diode temperature must be less than the manufacturer's maximum junction temperature rating for continuous operation. In part 2, the current is increased to 1.25 times the STC Isc for a second hour with the diode ambient still at 75 °C. The diode must still function as a diode after this second part. Many have questioned whether this test is severe enough for diodes used in modules that operate at higher temperatures, like those mounted on roof tops in high temperature locations.

As part of the TG 4 work, Shiradkar evaluated the expected diode operating temperature in multiple climates [34]. This work used modeling to estimate the maximum temperature that three specific types of didoes mounted in PV modules would ever see in two different mounting types (open rack and roof mount) in three different climate locations within the US (Denver, Miami and Phoenix). These temperatures were then compared to the temperatures that the same type diodes reached during the Bypass Diode Thermal Test. Table 7.5 shows the results for open-rack mounted modules. Table 7.6 shows the results for roof-mounted modules.

Table 7.5 Comparison of calculated field diode maximum temperature for open rack mount versus performance in bypass diode thermal test.

Location	Diode Type	Maximum Field Temp (°C)	Max Temp in MQT 18.1 with Isc (°C)
Denver	A	135	151
	B	148	165
	C	150	166
Phoenix	A	146	151
	B	160	165
	C	161	166
Miami	A	134	151
	B	147	165
	C	148	166

Table 7.6 Comparison of calculated field diode maximum temperature for roof mount versus performance in bypass diode thermal test.

Location	Diode Type	Maximum Field Temp (°C)	Max Temp in MQT 18.1 with Isc (°C)
Denver	A	159	151
	B	173	165
	C	174	166
Phoenix	A	172	151
	B	186	165
	C	187	166
Miami	A	157	151
	B	171	165
	C	172	166

The good news is that that none of the open-rack mounted didoes exceed the temperature they reached during the Bypass Diode Thermal Test (MQT 18.1) as shown in Table 7.5. None of these didoes should operate above the manufacture's maximum junction temperature rating for continuous operation. The bypass diode test in IEC 61215-2 is adequate for diode operation in open-rack mounting configurations.

The bad news is that all of the roof mounted diodes exceed the temperature they reached during the Bypass Diode Thermal Test (MQT 18.1) as shown in Table 7.6. Therefore, it is likely that these didoes will operate above the manufacturer's maximum junction temperature rating for continuous operation in a roof-mount configuration. The test we have in IEC 61215-2 is not adequate for testing diodes for roof mounting or possibly other higher temperature operations. This was addressed in a subsequent paper by Shiradkar [35], which looked at three different diodes in 16 climate zones around the world with two different mounting systems (open-rack and roof-mounted). Once again, the paper compared modeled outdoor data with results of the Bypass Diode Thermal Test (MQT 18.1). This time however, the results were expanded to higher temperatures (above 75 °) and higher current levels (above STC Isc). In this case, the paper identified four different scenarios:

- True positives where both the modeled temperatures and the measured MQT 18.1 temperatures were below the manufacturer's maximum junction temperature rating for continuous operation. These should be safe for use.
- True negatives (or inadequate designs) where both the modeled temperatures and the measured MQT 18.1 temperatures were above the manufacturer's maximum junction temperature rating for continuous operation. These would not be safe for use.
- False positives (an inadequate design qualified) where the measured MQT 18.1 temperatures were below the manufacturer's maximum junction temperature rating for continuous operation but the modeled temperatures were above the manufacturer's maximum junction temperature rating for continuous operation. So, these pass the test but are not safe for use.

- False negatives (adequate designs are disqualified by the test) where the measured MQT 18.1 temperatures were above the manufacturer's maximum junction temperature rating for continuous operation, but the modeled temperatures were below the manufacturer's maximum junction temperature rating for continuous operation. These fail the test but should be safe for use.

By varying the temperature and current used in the Bypass Diode Thermal test, the number of false positives and false negatives could be minimized. Based on this analysis, Shiradkar [35] recommended revising the Bypass Diode Thermal Test Conditions for higher temperature operations (roof-mounted and hotter climates) to:

- For part 1, a temperature of 100 °C with a current flow of 1.15 times STC Isc.
- For part 2, a temperature of 100 °C with a current flow of 1.4 times STC Isc.

These values have been proposed for implementation in the IEC 63126: "Guidelines for Qualifying PV Modules, Components and Materials for Operation at High Temperatures," which will be discussed in Chapter 10.

7.5 Task Group 5: Testing for UV, Temperature and Humidity

TG 5 is focused on the effects of UV radiation, which for many PV materials is coupled to temperature and humidity. Light (especially UV radiation) can cause changes in modules including discoloration, reduced adhesion at interfaces, embrittlement, light-induced degradation in the cells, and solarization of the glass. Often light stresses are applied to material coupons or a mini-module to reduce the cost of long-term exposures.

The leaders of TG 5 (Michael Koehl of Fraunhofer ISE, David Miller of NREL, Tsuyoshi Shioda of Mitsui Chemicals and Carol Chen of China National Electric Apparatus Research Institute) are trying to standardize the test conditions (including the UV sources, UV irradiance, temperature, and humidity) used to age specimens. They are addressing a number of failure mode issues including delamination and subsequent corrosion, so the scope of TG 5 overlaps with the scope of TG 3.

As was pointed out in Chapter 2, discoloration of the encapsulant has been one of the leading causes for module degradation over time. Much progress has been made in developing improved encapsulant formulations, but the qualification test (IEC 61215) only requires a short-term UV exposure, which is not long enough to assess whether a material (encapsulant or front sheet in particular) will discolor during operation and therefore result in reduced short-circuit current. Analysis of performance degradation from the PV literature by Jordan [36] determined that the degradation of cry-Si modules is still dominated by reduction in short-circuit current with encapsulant discoloration likely being a contributing factor. A UV exposure test for encapsulant and front-sheet loss of optical transmittance should then be a part of a module wear out testing program.

Dave Miller led a large international TG 5 group in conducting an experiment to determine whether accelerated UV testing could duplicate the change in optical transmittance observed in the field for the historical formulation of poly (ethylene-co-vinyl acetate) (EVA) [37, 38]. In the experiment, six different types of encapsulant were exposed including

(i) STR – A9918, the early material that discolored in the field (as described in Chapter 2), (ii) several EVA formulations that have been exposed in the field with minimal discoloration, (iii) an EVA formulation without UV stabilizers that is now being used in high efficiency cry-Si modules and (iv) a thermoplastic polyurethane (TPU) that is noted for discoloring quickly outdoors and, therefore, not used as a PV encapsulant. The experiment used two different sources of UV, a xenon lamp with a "right light" filter and a custom chamber with UVA-340 fluorescent lamps. As a reminder, Xe lamps produce a full spectrum of light from UV to IR but the UVA lamps only produce UV. The UVA samples were all exposed at 60 °C, while Xe exposures were performed at 45, 60, and 80 °C.

The encapsulant samples were laminated between two silica slides. This minimized both the UV absorbed in the superstrate and the amount of oxygen available for bleaching of any discoloration. The samples were exposed to UV at a level of $1.0\,W/m^2/nm$ (at 340 nm) at temperature for specified time periods, removed, measured for optical transmittance and then placed back in the UV chambers until they had been exposed for 180 days to a cumulative UV exposure level of $0.64\,GJ/m^2$. The results for the UVA fluorescent lamps are shown in Figure 7.9.

From this figure, it can be seen that the accelerated UV testing does duplicate the expected discoloration as the TPU discolors the quickest as we expected from field results while the EVA-A (STR A9918) also discolors significantly. None of the other EVA formulations had measureable loss in transmittance. So, in addition to confirming the STR finding that it is the additive in EVA that cause the discoloration, these results also confirm that we can use an accelerated UV test to evaluate encapsulants for discoloration. Now the question is what specifically should the test procedure look like?

The first thing to look at is whether the light source matters. The xenon and fluorescent light sources give different degradation rates for those materials that did degrade. While we

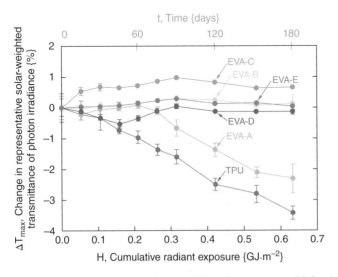

Figure 7.9 Change in transmittance with radiant exposure (H) for the coupons aged with UVA-340 lamps, with the chamber controlled at 60 °C [39]. *Source:* reprinted from Q-Lab/NIST Workshop with permission of David Miller.

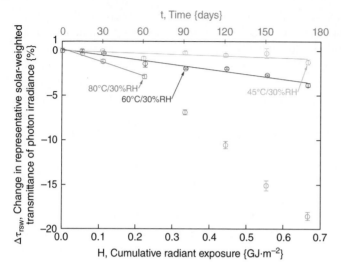

Figure 7.10 Comparison of change in transmittance with applied temperature for EVA-A using xenon source [38]. *Source:* reprinted from **Progress in PV** with the permission of the publisher, Wiley, Inc. and the author, David Miller of NREL.

can only conjecture why at this point (maybe photo-bleaching from the visible light in the xenon source or the different UV spectrum in the two), it does mean that when a standard is written it will have to select one of these sources as a primary method.

The second thing to look at is the temperature of the exposure. Figure 7.10 shows the change in transmittance as a function of temperature for EVA-A (STR A9918) using the xenon source. There is a significant temperature dependence that yields an estimated activation energy of ~60 kJ/mol. These results show that we will need to use higher temperatures if we want an accelerated test that can represent a significant fraction of the warrantied module lifetime.

Based on these results, a new IEC TS, IEC TS 62788-7-2 [40] has been developed. It calls for exposure rates similar to those used in the TG 5 experiment and allows for black-panel temperatures between 70 °C and 110 °C in steps of 10 °C. It is now up to the Working Group 2 to decide what document(s) should require a specific screening test for encapsulant discoloration and then to define the specifics of that test.

The same PVQAT TG 5 that worked on the UV-discoloration experiment also conducted an experiment to determine how UV exposure impacts the adhesion between the front glass and the encapsulant [41]. Glass/encapsulant/glass coupons were artificially aged using a variety of different steady-state conditions. The glass used was low iron and the EVA formulation was STR 15295P, a material with many years of successful field experience. These were exposed to UV under a variety of temperature and humidity conditions and then examined using a recently developed compressive shear test to quantify the strength of attachment at the EVA/glass interface.

At lower temperatures (~60 °C), little change in adhesion has been seen. However, at 80 °C there is significant degradation of the adhesion as shown in Figure 7.11. From this data, it is clear that UV exposure can reduce the adhesion between glass and EVA, in this case, by more than 50% from the initial value. However, the adhesion has reached an

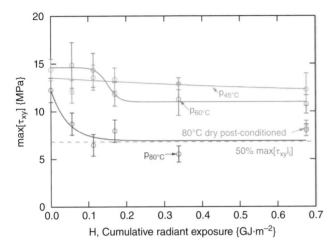

Figure 7.11 Xenon weathering performed at 80 °C and 30% relative humidity. *Source:* reprinted from NREL PVMRW with permission of David Miller.

asymptote and no further degradation is occurring. Even more interesting, the value it has leveled off at is in excess of 5 MPa, which is perfectly adequate to keep the module from delaminating.

To summarize the adhesion work, it is clear that UV exposure does have a negative impact on the adhesion between EVA and glass, but it has a limited effect that is not likely to be the cause of major module field failures. On the other hand, when selecting a new encapsulant or even front sheet for that matter, it would probably be a good idea to test the UV stability of adhesion as was done in this TG 5 experiment.

7.6   Task Group 6: Communications of Rating Information

TG 6 was originally set up to determine how the PV industry, particularly investors, wished to have a module durability rating system set up. The leaders of TG 6 (Sarah Kurtz of NREL and David Williams of CleanPath Ventures) organized several meetings to discuss this issue. There was no real consensus that came out of these meetings in terms of how the group wanted the rating information communicated, but there was a clear consensus that they did not want a system that rated modules on a particular scale, for example, A to F. So, rather than having a graded scale they proposed that any module rating system coming out of the PVQAT system should be set up to communicate that particular module types have passed the defined test protocol.

7.7   Task Group 7: Testing for Snow and Wind Load

TG 7 was focused on mechanical loading due to snow and wind, particularly those stresses that cause cell cracking. Early work of TG 7 led by Joerg Althaus of TUV resulted in the development of a test for inhomogeneous snow loads that is soon to be published as IEC 62938 [42]. In addition, a proposal was made to WG2 to redefine the minimum requirements for resistance to hail impacts. In IEC 61215-2: 2016, the definition of minimum requirements for hail testing was clarified. Since completion of these initial action items, the task group has been inactive.

7.8 Task Group 8: Testing for Thin-Film Modules

TG 8 is focused on the durability of thin-film PV modules. Subcommittees for thin films were established for the following areas:

- Micro-delaminations of device layers
- Shading effects in thin films
- Monolithic Integration
- Flexible packages
- Semiconductor junction degradation

The leaders of TG 8 (Originally Neelkanth Dhere of FSEC but later Michael Kempe of NREL and Masayoshi Takani of Solar Frontier) believe that the tests being developed for silicon modules can usually be applied to thin-film modules as has been done in IEC 61215-1 and -2. Representatives of the thin-film technologies have joined the other task groups to extend the discussions beyond the silicon-based module tests in order to ensure that these will be relevant to thin-film products as well. A separate thin-film test sequence is not being discussed, but discussions are ongoing to better understand failure modes that are unique to thin-film modules.

One example of this, is the hot spot test where monolithic modules are tested in a different way from wafer-based modules. The present hot spot test in IEC 61215-2 for monolithic modules is really only a safety test and does not address performance degradation. Timothy Silverman of NREL has prepared a draft IEC TS (IEC TS 63140) that provides a realistic partial shading test for thin-film modules [43].

7.9 Task Group 9: Testing for Concentrator Photovoltaic (CPV)

TG 9 is focused on durability of CPV modules. The leader of TG 8 (Nick Bosco of NREL) served as project leader for an IEC project to develop a new standard for testing CPV modules for longer-term survival against thermal-cycling stresses. IEC 62925 [44] has now been published. Since completion of this standard, the task group has been inactive.

7.10 Task Group 10: Testing for Connectors

TG 10 is focused on the durability of PV connectors. The group has examined evidence of connector failures and is planning to initiate tests on durability and reliability of contacts, especially their susceptibility to mechanical or corrosive failure mechanisms. Plastic materials in the bodies of the connectors need to retain their strength against UV and humidity exposure. Though there is already an IEC standard for PV Connectors, IEC 62852 [45], this really only covers safety not reliability or functionality. Therefore, the task group is working on developing qualification test procedures for connectors.

TG 10, under the leadership of Bryan Skarbek of First Solar, Alan Xu of Canadian Solar, and Yingnan Chen of China General Certification Center, has conducted a survey to prioritize the group's efforts, and is now working through a DFMEA to identify the most critical failure mechanisms observed for PV connectors.

7.11 Task Group 11: QA for PV Systems

TG 11 was created to support efforts to establish the IEC System for Certification to Standards Relating to Equipment for Use in Renewable Energy Applications (IECRE). Since much of IECRE's work is being done directly by IECRE using standards developed by IEC TC 82, the formation of this Task Group was mainly to provide a mechanism for communication between the separate efforts.

The leaders of TG 11 (Peter Hacke of NREL, Masaaki Yamamichi of AIST, and Yingnan Chen of China the General Certification Center) started a subgroup under Peter Hacke to work on the reliability of power electronics for use in PV.

7.12 Task Group 12: Soiling and Dust

TG 12 was created in 2014 to focus on understanding, quantifying, and mitigating losses of electricity generation from PV systems caused by soiling. TG 12 shares information on a variety of topics related to soiling including:

- Glass-coating technologies
- Economics of soiling losses and the tradeoffs from cleaning
- Methods for monitoring of soiling
- Soiling in severe locations
- Important field conditions for soiling – dust storms, wind velocity, etc.
- Health and safety concerns for dust storms and dust testing
- Modeling of the effects of soiling on PV systems; and
- Artificial soiling methods for use in the laboratory.

The purpose of the group's monthly webinars is to exchange relevant information, facilitate networking, and aid research, development and standards development related to soiling in PV modules and systems. The leaders of TG 12 (YuePeng Deng of First Solar, Mike Van Iseghem of EDF, Leonardo Micheli of NREL, Dave Miller of NREL and Lin Simpson of NREL) created the following four subtask groups:

1) A subgroup led by YuePeng Deng has focused its work on issues related to sensors and the monitoring of soiling. The group contributed to the standard IEC 61724-1 [46], which defines and specifies methods for quantifying the effects of soiling on PV systems.
2) A subgroup led by Lin Simpson has focused its efforts on finding solutions for the cleaning of soiling and accumulated contamination from PV modules. Specific topics

include developing the scientific basis for qualifying soiling of modules and methods to clean them off. This work includes looking at issues related to the economics of cost-effective cleaning.

3) A subgroup led by Dave Miller has been investigating anti-reflective and/or anti-soiling coatings. The group has been studying abrasion of the front glass of PV modules by the natural environment (the soiling process itself) or by processes used to clean the soiling off of the modules. One of the goals of the subgroup is to write a PV industry-specific standard related to abrasion.

4) A subtask group led by Leonardo Micheli has been analyzing the effects of soiling on the output power of PV systems. The group is attempting to develop a method to determine the soiling loss using power production data from actual PV power plants. The subgroup is trying to identify models that will predict typical soiling losses at sites located around the world.

7.13 Task Group 13: Cells

TG 13 is the newest group that was created to investigate challenges related to characterization of the performance and durability of PV cells in parallel with the newest IEC TC82 Working Group 8 on PV Cells. The leaders of TG 13 (Jin Hao of Jinko Solar and Sunit Tyagi of InSolare Energy) have begun the following three projects:

1) To update the efforts to write a standard for measuring the light-induced degradation in crystalline-silicon solar cells, IEC 63202-1 [47]
2) To develop a new test method for solar cell electrochemical capacitance-voltage measurement of P-N junction depth and turn it into an IEC standard.
3) To develop a new test method for measuring the performance of bifacial cry-Si cells.

References

1 PVQAT (2011). https://www.pvqat.org/events/1st_pvqat_forum.html (Accessed 29 August 2019).

2 PVQAT (2019). https://www.pvqat.org/index.html (Accessed 29 August 2019).

3 Jordan, D., Wohlgemuth, J., and Kurtz, S. (2012). Technology and Climate Trends in PV Module Degradation. NREL (National Renewable Energy Laboratory).

4 Norman, P., Sinicco, I., Eguchi, Y. et al. (2013). Proposal for a Guide for Quality Management Systems for PV Manufacturing: Supplemental Requirements to ISO 9001-2008. https://www.nrel.gov/docs/fy13osti/58940.pdf (Accessed 29 August 2019).

5 Eguchi, Y., Ramu, G., Lokanath, S.V. et al. (2014). Requirements for Quality Management System for PV Module Manufacturing. 40th IEEE PVSC in Colorado, USA (8–13 June 2014).

6 IEC TS 62941 (2016). Terrestrial photovoltaic (PV) modules – Guideline for increased confidence in PV module design qualification and type approval.

7 Ramu, G., Yamamichi, M., Zhou, W. et al. (2015). Updated Proposal for a Guide for Quality Management Systems for PV Manufacturing: Supplemental Requirements to ISO 9001-2008. https://www.nrel.gov/docs/fy15osti/63742.pdf (Accessed 29 August 2019).

8 Wohlgemuth, J.H., Cunningham, D.W., Amin, D. et al. (2008). Using Accelerated Tests and Field Data to Predict Module Reliability and Lifetime. 23rd EU PVSEC in Valencia, Spain (1–5 September 2008).

9 Herrmann, W. and Bogdanski, N. (2011). Outdoor Weathering of PV Modules – Effects of Various Climates and Comparison with Accelerated Laboratory Testing. 37th IEEE PVSC in Washington, USA (19–24 June 2011).

10 Kinsey, G., Meakin, D., Schmid, C. et al. (2013). Results of the First Round of the Photovoltaic Durability Initiative. 39th IEEE PVSC in Florida, USA (16–21 June 2013).

11 Wohlgemuth, J.H., Cunningham, D.W., Nguyen, A.M., and Miller, J. (2005). Long Term Reliability of PV Modules. 20th EU PVSEC in Barcelona, Spain (6–10 June 2005).

12 IEC TS 62782 (2016). Photovoltaic (PV) modules – Cyclic (dynamic) mechanical load testing.

13 Bosco, N., Silverman, T.J., Wohlgemuth, J. et al. (2013). Evaluation of Dynamic Mechanical Loading as an Accelerated Test Method for Ribbon Fatigue. 39th IEEE PVSC in Florida, USA (16–21 June 2013).

14 Dubey, R., Chattopadhyay, S., Kuthanazhi, V. et al. (2013). All-India Survey of Photovoltaic Module Degradation: 2013. National Centre for Photovoltaic Research and Education Indian Institute of Technology Bombay.

15 Jordan, D.C., Wohlgemuth, J., and Kurtz, S. (2012). Technology and climate trends in PV module degradation. 27th EU PVSEC in Frankfurt, Germany (24–28 September 2012).

16 Bosco, N., Silverman, T., and Kurtz, S. (2016). Climate specific thermomechanical fatigue of flat plate photovoltaic module solder joints. *Microelectronics Reliability* http://dx.doi.org/10.1016/j.microrel.2016.03.024.

17 King, D.L., Boyson, W.E., and Kratochvil, J.A. (2004). Photovoltaic array performance model. Sandia National Laboratories. SAND2004-3535

18 Norris, K. and Landzberg, A. (1969). Reliability of controlled collapse interconnections. *IBM Journal of Research and Development* 13: 266–271.

19 Bosco, N., Silverman, T., and Kurtz, S. (2016). The influence of PV module materials and design on solder joint thermal fatigue durability. *IEEE Journal of Photovoltaics* 6 (6): 1407–1412.

20 IEC 62892 (2019). Extended thermal cycling of PV modules – Test procedure.

21 TamizhMani, G. (2014). Reliability Evaluation of PV Power Plants: Input: Data for Warranty, Bankability and Energy Estimation. NREL PVMRW.

22 Wohlgemuth, J., Silverman, T., Miller, D.C. et al. (2015). Evaluation of PV Module Field Performance. 42nd IEEE PVSC in New Orleans, USA (14–19 June 2015).

23 Bosco, N. (2015). Moving the PV Industry to a Quantitative Adhesion Test Method. 3rd Atlas/NIST Workshop on PV Materials Durability in Maryland, USA (8–9 December 2015).

24 Wohlgemuth, J.H., Hacke, P., Bosco, N. et al. (2016). Assessing the Causes of Encapsulant Delamination in PV Modules. 43rd IEEE PVSC in Oregon, USA (5–10 June 2016).

25 Miller, D.C., Annigoni, E., Ballion, A. et al. (2016). Degradation in PV Encapsulant Adhesion: An Interlaboratory Study Towards a Climate-Specific Test. 43rd IEEE PVSC in Oregon, USA (5–10 June 2016).

26 IEC TS 62804-1 (2015). Photovoltaic (PV) modules – Test methods for the detection of potential-induced degradation – Part 1: Crystalline silicon.

27 Hacke, P., Smith, R., Terwilliger, K. et al. (2013). Development of an IEC test for crystalline silicon modules to qualify their resistance to system voltage stress. *Progress in photovoltaic: Research and Applications* https://doi.org/10.1002/pip.2434.

28 Hoffmann, S. and Koehl, M. (2012). Influence of Humidity and Temperature on the Potential Induced Degradation. 27th EU PVSEC in Frankfurt, Germany (24–28 September 2012).

29 Hacke, P. (2014). Testing modules for potential-induced degradation – a status update of IEC 62804. NREL PVMRW.

30 Hacke, P. (2017). PVQAT TG3: Proposed PID pass-fail requirement for amendment to IEC 61215, other TG3 status and combined stress testing. NREL PVRW.

31 Eric, J.S., Paul Brooker, R., Shiradkar, N.S. et al. (2016). Manufacturing metrology of C-Si module reliability and durability part III – module manufacturing. *Renewable and Sustainable Energy Reviews* 59: 992–1016. http://dx.doi.org/10.1016/j.rser.2015.12.215.

32 IEC TS 62916 (2017). Bypass diode electrostatic discharge susceptibility testing for photovoltaic modules.

33 IEC 62979 (2017). Bypass diode thermal runaway test.

34 Shiradkar, N., Gade, V., Schneller, E.J., and Sundaram, K.B. (2015). Revising the Bypass Diode Thermal Test in IEC 61215 Standard to Accommodate Effects of Climatic Conditions and Module Mounting Configurations. 42nd IEEE PVSC in New Orleans, USA (14–19 June 2015).

35 Shiradkar, N.S., Gade, V.S., Schneller, E.J., and Sundaram, K.B. (2018). Revising the Bypass Diode Test to Incorporate the Effects of Photovoltaic Module Mounting Configuration and Climate of Deployment. WCPEC-7.

36 Jordan, D.C., Wohlgemuth, J.H., and Kurtz, S.R. (2012). Technology and Climate Trends in PV Module Degradation. 27th EU PVSEC in Frankfurt, Germany (24–28 September 2012).

37 Miller, D.C., Annigoni, E., Ballion, A. et al. (2015). Degradation in PV Encapsulation Transmittance: An Interlaboratory Study Towards a Climate-Specific Test. 42nd IEEE PVSC in New Orleans, USA (14–19 June 2015).

38 Miller, D.C., Bokria, J.G., Burns, D.M. et al. (2019). Degradation in PV encapsulation transmittance: results of the first PVQAT TG5 study. *Progress in Photovoltaics: Research and Applications* https://doi.org/10.1002/pip.3103.

39 Miller, D.C., Bokria, J.G., Burns, D.M. et al. (2018). Degradation in PV Encapsulation Transmittance: Results of the First PVQAT TG5 Study. Q-Lab/NIST Service Life Prediction Symposium in Colorado, USA (18–24 March 2018).

40 IEC 62788-7-2 (2017). Measurement procedures for materials used in photovoltaic modules – Part 7-2: Environmental exposures – Accelerated weathering tests of polymeric materials.

41 Fowler, S., Xiaohong, G., Miller, D., and Phillips, N. (2017). PVQAT TG5: UV Weathering Standards Development Within the PV Industry. NREL PVMRW.

42 IEC 62938 (To be published in 2020): Non-uniform snow load testing for photovoltaic (PV) modules.

43 IEC T 63140 (To be published) – Photovoltaic (PV) modules – Partial shade endurance testing for monolithically integrated products.

44 IEC 62925 (2016). Concentrator photovoltaic (CPV) modules – Thermal cycling test to differentiate increased thermal fatigue durability.

45 IEC 62852 (2014). Connectors for DC-application in photovoltaic systems – Safety requirements and tests.

46 IEC 61724-1 (2017). Photovoltaic system performance – Part 1: Monitoring.

47 IEC 63202-1 (2019). Photovoltaic cells – Part 1: Measurement of light-induced degradation of crystalline silicon solar cells.

8

Conformity Assessment and IECRE

Conformity Assessment refers to activities that are undertaken to determine that products, services, or even systems meet the requirements of their specification and/or relevant standards. You can see how important this would be for photovoltaics (PV) as they require huge upfront investments and then the economic return is based on them performing to specification for many years. Most of this book so far has detailed the types of extended accelerated testing and factory quality control necessary for PV modules to meet specifications and warranties. However, there are two additional requirements that must be met to achieve the type of conformity assessment needed in PV. These will be discussed in the next two sections. The first section discusses development of conformity assessment systems for PV products, in this case, mostly PV modules although some PV products like PV lanterns have also been covered. The second section discusses the extension of conformity assessment from the product to the PV system deployed in the field and how this required the creation of a whole new Conformity Assessment organization at the International Electrotechnical Commission (IEC).

8.1 Module Conformity Assessment – PowerMark, IECQ, PVGAP, and IECEE

The simplest type of conformity assessment is for a specific product which in the case of PV includes the module. Clearly some sort of assurance that the modules meet their specifications and the requirements of international standards is important for those buying and using modules. There are really three types of conformity assessment systems based on who is performing the assessment.

1) First-party conformity assessment is where the company manufacturing the product performs the assessment including its use of a quality management system (QMS) and product accelerated testing like IEC 61215 for qualification and IEC 61730 for safety. Some manufacturers will even provide a Supplier Certificate of Conformity. In many ways, every supplier that claims to meet their product specifications and specific international standards is claiming conformity assessment. This is the least expensive type of conformity assessment but also provides the least assurance to the customer.

2) Second-party conformity assessment is where the purchaser/user of the product performs the assessment based on their perceived requirements. The Jet Propulsion Laboratory (JPL) Block Buy System discussed in Chapter 4 was one of the first Second-Party Conformity Assessment systems for PV. JPL provided specifications for the modules, the accelerated stress tests to be performed on them and approved the quality system under which they were built. Typically, large, important customers use second-party conformity assessment systems. There are many examples of this in PV where the module manufacturer had to meet the customer's product specification, accelerated test protocol and QMS audit. Such systems were described in Section 6.4 and the problems with this type of customer system were detailed in Section 6.5.

3) Third-party conformity assessment is where the assessment is performed by an independent body that is not connected with either the manufacturer/supplier or with the customer. In this case, the conformity assessment is usually done to international standards and when complete is called a certification. Third-party conformity assessment is more reliable than first party since it is an independent look at the manufacturer's specifications, testing and quality management. Since a third-party conformity assessment is also independent from the customer it has the potential to be accepted by many different customers, thereby ultimately being more cost effective. One well-known example of such a third-party system is the Underwriter's Laboratory system for safety certification.

Once IEC 61215 was published some module purchasers, particularly those purchasing larger numbers of modules for utility-scale applications, began requiring third-party module certification to IEC 61215. To provide this third-party service, the European Solar Test Installation (ESTI) located at the Joint Research Center (JRC) at the Ispra site in Italy began testing PV modules to IEC 61215. To help serve the PV industry, ESTI obtained third-party accredited to perform PV-module measurements and calibration as well as for testing to IEC 61215. Many European customers then began requiring third-party certification by a laboratory accredited to perform IEC 61215. At this time, there were no laboratories in the US that were third party accredited to perform IEC 61215. US module manufacturers had to send their modules to ESTI in Italy for IEC 61215 certification. Having only one laboratory to use, and that one being on a different continent, resulted in long waits for certification, handicapping the US PV industry.

To address the laboratory issue, as well as to begin to organize a third-party conformity assessment system in the US, the Department of Energy (DOE) established a program led by Carl Osterwald at the National Renewable Energy Laboratory (NREL) to develop a plan for PV-module certification and PV-laboratory accreditation. With the help of a group from Arizona State University led by Bob Hammond, and a large committee from the US PV industry, a plan for such a program was developed and published [1]. It is of interest to review this document as it provides a look at what is required for a product certification and laboratory accreditation system. The major part of the published handbook consisted of the following documents.

8.1.1 PV-1: "Criteria for a Model Quality System for Laboratories Engaged in Testing PV Modules"

PV-1 provided the basis for designing and implementing quality systems in laboratories and established the requirements for recognizing the competence of laboratories via accreditation. PV-1 was based on ISO/IEC Guide 25 [2] for laboratories, but it also provided the specifics necessary for PV-module testing laboratories. The implementation criteria committee selected the American Association for Laboratory Accreditation (A2LA) to perform the laboratory accreditations.

Document PV-1.1: "Calibration, Traceability, and Statistical Requirements of Testing by Accredited Laboratories in Support of the Photovoltaic Module Certification Program" was added to PV-1 to provide a minimum set of requirements for PV-module calibration, calibration traceability and testing to which a laboratory should comply. PV-1.1 was to be used as a model for the development of a laboratory's quality system, and was also intended to be used in the subsequent assessment of a laboratory's quality program.

8.1.2 PV-2: Model for a Third-Party Certification and Labeling Program for PV Modules

This document provided a model of a third-party certification/conformity assessment system. It used the product standards given in PV-3, requiring testing to those standards and then assessment of the manufacturer's quality management system (QMS) utilized to make that product. Subsequent requirements included continued factory surveillance and periodic product re-testing. PV-2 was designed to be consistent with international standards on certification and conformity assessment including ISO/IEC Guide 28 [3].

8.1.3 PV-3: Testing Requirements for a Certification and Labeling Program for PV Modules

PV-3 defined the testing requirements for PV-module qualification and for electrical performance measurements that should be used in support of a PV-module certification and labeling program modeled in PV-2. PV-3 specified that module designs be tested to Institute of Electrical and Electronics Engineers (IEEE) Standard 1262 [4]. PV-3 contained a list of required equipment for testing PV modules as well as the necessary staffing including the requirements for their qualification and training.

While IEC 61215 had already been published by the time the handbook was published, IEEE 1262 was selected over it for several reasons including the fact that the IEEE document was written for all PV technologies while IEC 61215 only covered crystalline silicon-based modules. As we will see later, selection of IEEE 1262 instead of IEC 61215 proved to be a mistake.

8.1.4 PV-4: Operational Procedures Manual for the Certification Body of the PV Module Certification Program

This document provided the procedures and processes required for the day-to-day operation of such a certification/conformity assessment program. It gave rules that were to be followed by any organization that wanted to set up such a system, as well as detailing what a module manufacturer would have to do, if they decided to participate. Document PV-4 was the principal guide for managing and administering a module certification program. It was designed to be consistent with PV-2 and ISO/IEC Guide 40 [5].

8.1.5 PV-5: Application and Certification Procedures for the PV Module Certification Program

PV-5 provided the procedures and processes required of PV-module manufacturers who would choose to obtain certification/conformity assessment on one or more of their module types.

This report led to the creation of PowerMark Corporation (PMC) whose charter was implementation of the recommendations from the handbook.

PowerMark Corporation: PMC was formed as a non-profit corporation to implement the programs developed in the Handbook [1]. Steve Chalmers served as the initial Executive Director of PowerMark. The Board of Directors included members from a number of prominent PV organizations of the day including; Solarex Corporation, ASE Americas, Arco Solar, NREL, Utility Photovoltaic Group, and Arizona State University. The main goal of PMC was to assist the industry in producing quality PV products. PMC's aim was to set up a laboratory accreditation program and a module certification/conformity assessment program.

The laboratory accreditation program was immediately successful. The Photovoltaic Testing Laboratory at Arizona State University (ASU) under Bob Hammond applied to PowerMark for accreditation to test PV modules. The ASU efforts were successful as they were granted accreditation through A2LA and PowerMark. Even more importantly, module manufacturers began having their modules tested at ASU and those test results were accepted around the world. Eventually, PMC accredited additional test laboratories like Florida Solar Energy Center for PV module power measurements.

The PMC efforts on product certification were not as successful. The PV industry was already transitioning from national standards (like IEEE 1262) to the international standards like IEC 61215. PMC did recognize this and modified their PV-3 document to include requirements for both IEEE 1262 and IEC 61215. However, by not removing IEEE 1262, they probably sacrificed their chance to have at least one PV-module manufacturer certify modules under the system. IEEE 1262 included a dynamic mechanical load test requiring 10 000 cycles at 1440 Pa. This test had been part of the JPL Block IV test procedure and was not particularly difficult for the small modules available at that time to pass. However, as modules and cells got much larger and cells got much thinner, this test became too severe. Some of the new, larger Solarex modules failed this test with multiple open circuits but have now survived in the field for more than 30 years. If PowerMark had only required qualification to IEC 61215, Solarex would probably have proceeded to get PowerMark certification. As it turned out, the IECQ system came along before PMC made the change to just IEC 61215 and Solarex went the IECQ route as will be discussed in the next subsection.

As an aside, PMC still exists today, but not as a product or laboratory conformity assessment organization. Since the late 1990s, PowerMark has managed the United States Technical Advisory Group (USTAG) for IEC TC82. In this capacity, PMC manages the US inputs into the IEC PV efforts.

IECQ: "IECQ is a worldwide approval and certification system covering the supply of electronic components and associated materials and assemblies and processes. IECQ uses quality assessment specifications that are based on International Standards that have been prepared by the International Electrotechnical Commission (IEC)" [6]. Founded in 1982, IECQ was the first of the IEC conformity assessment systems to become operational.

IECQ set up a conformity assessment system for PV modules soon after IEC 61215 was published, when it became obvious that IEC 61215 was going to be a popular and important qualification scheme. The IECQ system for PV modules and components included product certification under IEC 61215 (or other relevant IEC standards) and IECQ certification of the manufacture's quality management system to ISO 9000. The system also had requirements for periodic product retests and inspections. PV products qualified under the system used a document called a "Blank Detailed Specification" that defined what tests and what testing frequency was required to maintain certification. Those meeting the requirements could then display the IECQ Mark that customers should be able to distinguish as meaning quality products.

Initially, several PV module manufacturers including Solarex and Unisolar obtained IECQ certification on their PV modules as they felt that IEC international standards were more important than localized country standards. However, over time, neither company found that having IECQ module certification gave them much of an advantage in the commercial marketplace. Within a few years, both companies dropped their IECQ certifications as they were expensive to maintain and put additional burdens on a manufacturer when they wanted to make changes to the products. There doesn't seem to be any records of other module manufacturer obtaining an IECQ certification for a PV module type.

The big question then is "why was the IECQ system for PV modules not successful?" The simple explanation is that customers just did not require or seem to value the IECQ certification on a PV module. Of course, that leads to the bigger question of why customers didn't value this conformity assessment system. The answer appears to be that the IECQ system really didn't add anything to product quality. The customers already purchased product qualified to IEC 61215 (or IEC 61646 if it were thin film) that were built in an ISO 9000 certified factory. The system was supposed to include post-certification audits, but there was no definition of how these audits were to be performed or how often they were required. Some certification bodies did the audits and others did not, but customers couldn't tell from the certification whether audits were being performed or not. So, it appeared to customers that the IECQ system did not add any additional technical value to the modules (no additional testing and no additional factory audit requirements).

The IECQ system for PV modules remained in existence until it was subsequently transferred to the IEC System of Conformity Assessment Schemes for Electrotechnical Equipment and Components (IECEE) in 2005. The IECEE PV system will be discussed in a subsequent subsection.

The Global Approval Programme for Photovoltaics (PV GAP): PV GAP was incorporated in Switzerland as a not-for-profit organization in 1996 by Peter Varadi (one of the cofounders of Solarex Corporation) and Marcus Real, Convenor of WG3 on PV systems. PV GAP's purpose was to establish a quality system for PV products with defined requirements as to how a product could obtain this PV Quality Mark [7]. PV GAP was particularly interested in assuring the quality of PV products sold into developing nations and one of its major supporters was the World Bank. The products to be certified under the PV GAP program included PV modules, charge controllers, inverter, some PV appliances like lamps and small PV power systems. PV GAP used IEC standards when available and created what they called Provisional International Specifications (PVRS) when there was no IEC standard available. As in the other certification/conformity assessment systems, earning the PV GAP Quality Mark required:

- Product tested to an international standard (or PVRS) by an accredited testing laboratory.
- Product manufactured in an ISO 9001 factory certified by an ISO registrar.
- Periodic unannounced inspections of the manufacturing facility.
- Selection of products for retest during unannounced inspections.

To simplify the management of the PV GAP system, IEC provided the secretarial services under the IECQ system.

A number of manufacturers qualified their product under the PV GAP system. In many cases, the World Bank required the use of PV GAP marked PV products in projects they funded. The use of such a quality system to produce quality products in countries like India and China had a significant impact on the acceptance of PV in those countries.

At about the same time that the PV system was transferred from IECQ to IECEE, PV GAP transferred its system to IECEE and stopped operating. The story then shifts to IECEE.

IECEE, the IEC System for Conformity Assessment Schemes for Electrotechnical Equipment and Components: IECEE "is a multilateral certification system based on IEC International Standards. Its Members use the principle of mutual recognition (reciprocal acceptance) of test results to obtain certification or approval at national levels around the world [8]." The IECEE system deals with more complex products than the IECQ system as it covers electrotechnical equipment rather than electronic components.

The IECEE system also includes a system for accrediting test laboratories. IECEE qualifies National Certification Bodies (NCBs) that are responsible for recognizing and issuing CB Test Reports and Certificates. The NCBs employ test laboratories, known as CB Testing Laboratories (CBTLs) to perform the tests in compliance with IEC International Standards. PVs first experience with IECEE was in adding the testing for module qualification and safety (IEC 61215 and IEC 61730) to the list of standards that an IECEE test laboratory could be accredited to.

In 2005, PV product certifications were moved from the IECQ system to the IECEE system. At the same time, PV GAP also transferred its system to IECEE. In 2019, the IECEE website lists more than 200 PV products that have been certified under the IECEE system. Most of these are PV modules. The IECEE PV program is described in reference [9]. This IECEE document claims that by using the IECEE system you can save time and money by:

- Reducing the number of steps required for certification.
- Eliminating duplicate testing.
- Only requiring testing of one sample type among similar families of products.
- Only requiring specific retests for product modifications.
- Providing acceptance by regulators, retailers, buyers and vendors in many countries.

Many of these features are already built into the IEC standards so are not really unique to IECEE.

On the downside, IECEE has not adopted IEC TS 62941 [10] which could serve as a basis for uniform factory auditing requirements. The different CBs within the IECEE system have implemented factory audits in their own way with methods and frequency varying around the world. Since there has been no agreement within the IECEE system on the auditing, many NCBs only accept certifications issued under their own system so the "one test, one certificate" slogan of IECEE has really not worked well in PV.

8.2   IECRE – Conformity Assessment for PV Systems

One thing that has become clear over the years is that a PV system can operate poorly even if it is built using the best quality components that are available. For a PV system to work well for its intended life, it must be designed, installed and maintained correctly. If any of these steps are not done correctly, the system itself may not function well or some of the components may fail prematurely. For example, modules can be damaged during installation so that cells break. Often broken cells don't result in power loss until years later as the stresses related to thermal cycling and wind loading cause the cracks to open up.

What is needed then, is a conformity-assessment system for the PV power plants that actually produce the electricity in the field. When this issue was first discussed in the early 2010–2012 timeframe, IEC did not have a conformity-assessment system that could cover this. The three IEC conformity-assessment systems in existence at the time were:

1) IECQ that certified electronic components.
2) IECEE that certified electronic equipment.
3) IECEx that certified equipment for use in explosive atmospheres.

Actually, the PV industry was not the first group to realize that a conformity-assessment system for power plants was needed and that IEC did not have an appropriate organization for this. TC 88, the IEC Technical Committee on Wind Energy, led by Sandy Butterfield of the US, wished to have standards written for certifying the performance and safety of installed wind turbines. IEC Central Office replied that certifying wind turbines was a conformity-assessment responsibility, but that there was no IEC conformity-assessment system that could do this. So TC88 went ahead and wrote the Quality Management documents for certifying wind energy systems. (It is often easier to ask forgiveness for doing something against the IEC rules than it is to get permission beforehand to do it.) Once these documents were completed, TC88 went back to IEC to lobby them to set up a new Conformity Assessment System for wind energy systems or wind farms.

In the meantime, Sandy Butterfield had contacted representatives of IEC TC82; including Howard Barikomo, the Secretary of TC 82 at the time, to see if the PV industry was interested in developing an IEC Conformity Assessment System that could also cover PV power plants. Howard and Sandy rallied the PV and wind industries as well as the Marine Energy Industry (TC 114) to petition IEC to establish a new Conformity Assessment System that could cover renewable energy power plants. These groups met with IEC management during an IEC General Meeting in October, 2012 in Oslo, Norway. During that meeting, the IEC management representatives finally realized that their three, existing conformity-assessment systems were not well suited to extend conformity assessment to renewable energy power plants. They agreed to set up a new conformity-assessment system specifically for certifying renewal energy power plants.

In 2013, the IEC Conformity Assessment Board (CAB) approved the creation of IECRE (IEC System for Certification to Standards Relating to Equipment for Use in Renewable Energy Applications) with three sectors; wind, PV and marine energy. Additional renewable energy sectors like solar thermal would be welcome to join in the future if they wished. An international team with members from the three IEC Technical Committees (TC82 on Solar Photovoltaic Energy Systems, TC88 on Wind Turbines, and TC114 on Marine Energy) with assistance from the IEC central office wrote the Basic Rules for the IECRE system. These were approved by the CAB in 2014 so the system was up and running.

IECRE conformity assessments will be performed by independent third parties (called certifying bodies), and will be based on inspections and test results provided by accredited laboratories. The assessments are to cover both factory aspects including certification of products like modules and inverters as well auditing of the manufacturer's QMS and field aspects including system design, installation, commissioning, operation, and maintenance [11].

The next step was for each of the three technical areas to prepare their own detailed assessment system. The Solar PV Operating Management Committee (PV-OMC) under Adrian Häring of SolarEdge from Germany (the Chairman of the PV-OMC) began the effort to develop Rules of Procedure for the PV sector [12]. There were three parts required for putting a system in place. These included (i) establishing the Rules of Procedure and the operating documents for the PV sector, (ii) determining what standards are to be followed and (iii) defining how certification bodies are to perform the audits/inspections and how they are to be accredited for this.

IECRE assessments should be done to International Standards, so one of the first steps in establishing the system for PV was to decide what standards to use. Standards are necessary for (i) products, (ii) quality management systems (QMS), (iii) PV system design, (iv) PV system installation and commissioning and (v) PV system operation and maintenance. Let's take a look at the status of each of these areas in terms of the standards that were available at that time (2014) and what standards had to be developed before the IECRE system could be functional.

1) PV product standards: A number of PV product standards were available for use including:
   - For PV modules – IEC 61215 on Qualification and IEC 61730 on Safety.
   - For PV Inverters – IEC 62891 on performance [13] and IEC 62109-1 [14] and IEC 62109-2 [15] on safety. There was also an old design qualification standard (IEC 62093 [16]) but it was not in use by many inverter manufacturers.
   - For Balance of Systems components – Standards for balance-of-system (BOS) components were not as well developed. There was a new standard for PV connectors (IEC 62852 [17]) and one on fuses for PV (IEC 60269-6 [18]. There were also some additional IEC standards that had been written by different Technical Committees (not TC82) for combiner boxes and disconnects but it wasn't clear whether these could be useful for IECRE or not.

So, even in the more developed product area there was still work to be done in terms of writing new standards (and updating some of the older ones) for the IECRE system.

2) <u>QMS for products</u>: As IECRE began, many PV module manufacturers were complying with the requirements of ISO 9001 so this was considered the initial baseline for assessment under IECRE. However, by this time the effort to develop a PV-module specific set of Quality Management requirements was already underway in PVQAT as was discussed in the previous chapter. Similar QMS standards would eventually be required for inverters and maybe also for some of the other BOS components.

3) <u>System Design</u>: IEC had already published IEC 62548 [19] on PV system design which could serve as a baseline for IECRE. Additional standards would be required including completion of a standard then in development to address the specific issues of system design for PV power plants as it was assumed that most IECRE assessments would be performed on larger utility scale systems.

4) <u>Installation and Commissioning</u>: At this time, there were no IEC installation standards available and very few under development. IEC 62446 [20-23] did provide guidance on requirements for system documentation, commissioning tests and inspection. There was also one out-of-date standard (IEC 61829 [24] on measuring the output power of a PV array, that badly needed updating.

5) <u>Operation and Maintenance</u>: IEC 61724 [25-27] contained guidelines for some aspects of system performance monitoring, but it was not very complete. There were really no IEC documents available at that time that addressed the maintenance of PV systems. A new series of documents would be necessary to provide guidance on operating and maintaining PV systems.

So, in addition to developing the Rules of Procedure, a large number of PV standards had to be improved or developed for the IECRE effort. George Kelly of Sunset Technology, the new Secretary of IEC TC2 and of the USNC/IECRE and Sarah Kurtz of NREL led the efforts to develop the standards necessary for the PV sector in IECRE. Since the initiation of the IECRE effort in 2013, a number of relevant IEC standards have been published including:

1) PV product standards:
   a) IEC 61215 on module qualification and IEC 61730 on module safety have been updated.
   b) IEC TS 62915 [28] defining module retest requirements for IEC 61215 and IEC 61730.
   c) IEC 62894 [29] on inverter datasheet and nameplate requirements.
   d) IEC 62817 [30] on design qualification for solar trackers.
2) QMS for products:
   a) IEC TS 62941 [10] the Guideline for QMS used to fabricate PV modules.
   b) There is an ongoing effort in TC82 to develop IEC 63157 entitled "Guidelines for effective quality assurance of power conversion equipment for photovoltaic systems" but publication is not expected until 2020 at the earliest.
3) System Design:
   a) IEC 62548 [19] on array design was updated in 2016.
   b) IEC 62738 [31] on array design for utility scale plants was published in 2018.

4) Installation and Commissioning:
   a) IEC 62446 was reissued as IEC 62446-1 [21] in 2016 providing guidance on documentation, commissioning tests and inspection.
   b) IEC 62446-3 [22] was issued in 2017 to provide an IR method for inspection of PV arrays.
   c) IEC 61529 [24] on measuring the performance of PV arrays was updated in 2015.
   d) IEC TS 63049 [32] providing a guideline for quality assurance in system installation, operation and maintenance was published in 2017.
5) Operation and Maintenance:
   a) IEC 61724 [25-27] was split into three parts with IEC 61724-1 covering monitoring, IEC 61724-2 covering capacity evaluation and IEC 61724-3 covering energy evaluation.
   b) IEC TS 63049 [32] mentioned above under Installation also provides guidelines for operation and maintenance.

This expanded set of standards was adequate for IECRE to begin certifying PV power plants.

The PV Management Committee (PV-OMC) worked on development of the operational documents (ODs) during this same time period. These documents define the requirements for certification. Each system certification project that takes place will be defined by three classes with the following subsets:

1) Lifecycle Class – Design, commissioning, or operation
2) Operator Class – Utility, commercial, residential, or aggregate
3) Locator Class – Ground, roof, or BIPV

Each type of certification is defined in a specific document. Table 8.1 gives a list of those ODs that have been completed and approved as of 2019. The PV-OMC anticipates preparation of more ODs as the need arises. Additional ODs that are under development or planned are given in Table 8.2 [33, 34].

The third requirement for starting the IECRE system was to identify the certifying bodies and test laboratories that will perform the inspections and audits. The other three parts of IEC Conformity Assessment already have a system for mutual recognition of

Table 8.1 Published IECRE PV system certification documents.

OD#	OD certification title
401	Conditional PV plant certificate
401-1	Conditional PV project certificate-supplement: site regulation, civil and construction work
402	Annual PV plant performance certificate
404	PV Plant operational status Assessment
405-1	IECRE Quality System Requirements for Manufacturers – Part 1: Requirements for certification of a quality system for PV module manufacturing
410-1	IECRE Quality System Requirements for PV Plant Installation and Maintenance – Part 1: Requirements for certification

Table 8.2 Planned IECRE PV system certification documents [33, 34].

OD#	Planned OD certification title
403	PV Plant Design Qualification Certificate
403-1	PV Site Qualification Certificate
403-2	PV Power Block Design Qualification Certificate
407	PV Parameters and Definitions
408-1	Procedures for the Issuing of IECRE-PV Certificates of Conformity
408-2	Procedures for the Issuing of IECRE-PV Test Reports
408-3	Procedures for the Issuing of IECRE-PV Quality Assessment Reports
408-4	Procedures for the Issuing of IECRE-PV Peer Assessment Reports
409	PV Decommissioning Certificate

test laboratories. The basis for establishing mutual recognition includes accreditation (assuring that an organization has competency in the standards being used in the system) combined with regular peer assessments by a committee of the other laboratories in the system to monitor performance and consistency. IECRE was able to use this same system for identifying the certification bodies (CBs) and test laboratories. CBs and laboratories that were already involved in testing for the PV industry applied for and were accepted into the IECRE Conformity Assessment System. In 2017, seven RECBs (Certification bodies for renewable energy) and eight REIBs (Inspection bodies for renewable energy) were approved for participation in the IECRE PV Sector [35].

As of early 2017, IECRE was ready to begin certifying PV power plants. There was no rush of customers to have their power plants assessed and certified. Indeed, as of mid-2019 there are still no PV power plants certified under IECRE, although the wind turbine part of IECRE has been issuing certificates since 2016 and as of early 2019, had issued 33 certificates around the world [36]. The PV branch has actually issued one certificate. It was issued in 2018 to the First Solar Module factory under IEC TS 62941 for module manufacturing quality. So, no PV power plants have been certified under IECRE yet.

To facilitate adoption of the PV IECRE system in the US, NREL issued a request for proposal and selected three contractors to receive financial support to have their PV power plants evaluated under the IECRE PV system [37]. The three plants evaluated in the program are identified below.

1) Ground mounted fixed tilt 75 MW utility array in Florida that was inspected by UL.
2) Single-axis tracking, 25 MW utility array in South Carolina that was inspected by Intertek.
3) Pole-mounted fixed tilt 1 MW carport array in California that was inspected by TUV.

One of the major technical conclusions of the study was that "Almost everything works well, conforms to standards, and generally looks nice" [37]. On the other hand, the effort identified a few issues with the standards and with the Rules of Procedure as they are now written. These are all things that should be corrected as soon as possible. The biggest issue identified was the lack of clarity about the relationship between the inspection body and the certification body. More than six months after completion of the onsite inspections, none of the three inspection organizations has submitted a report nor have any of the three systems yet received an IECRE certificate.

So why hasn't the IECRE system for certifying PV power plants gotten off the ground quickly? Why does there seem to be little enthusiasm among system owners to use the IECRE conformity-assessment system? There are a number of likely reasons:

- Change is always hard, especially when it costs money.
- The system that has been in use for large power plants (hiring a PV knowledgeable engineering firm and requiring extended product testing and factory audits which is really a second party system) has proven quite successful. Most PV systems have worked well.
- Much of the work spearheaded by PVQAT (see Chapter 7) has not yet been incorporated into the IECRE system. These include:
  - Major improvements in qualification testing including PID and Dynamic Mechanical Load to test for cell breakage have not yet been incorporated into IEC 61215. Those changes are being including in the next edition of IEC 61215 which is still under development (see Chapter 10).
  - Major improvements in safety testing, including extended UV testing of encapsulants, frontsheets and backsheets have not yet been incorporated into IEC 61730. Those changes are being included in an amendment now working its way through the IEC approval system (See Chapter 10).
  - The IERE product certification does not include any of the extended accelerated stress tests often required today for the modules going into larger PV systems. As indicated in Chapter 6 this extended testing, which often is not grounded completely in science, does include some valuable testing like extended thermal cycling. Until the IEC can finalize standards around the extended testing being developed under PVQAT, the IECRE certificate is not likely to be seen as equivalent to people's own system and therefore inadequate for long-term survival of the PV modules.
- Since the IECEE product certifications have not been accepted worldwide by the different certification bodies, there may be some reluctance to accept them as part of the IECRE power plant certifications.
- While having a product recognized globally may save testing cost and time, it may be less clear that a global system of conformity assessment for PV power plants will actually be able to reduce overall testing and inspections. Each power plant would still have to be evaluated on its own design merits and inspected for compliance.

The jury is certainly still out on the IECRE PV power plant certification system. The system must generate business soon or it will not be able to survive, especially because of the high cost of maintaining the administrative offices.

References

1 Osterwald, C.R., Hammond, R.L., Wood, B.D. et al. (1996). Photovoltaic Module Certification/Laboratory Accreditation Criteria Development: Implementation Handbook. NREL/TP-412-21291.

2 ISO/IEC Guide 25 (1990). General requirements for the competence of calibration and testing laboratories.

3 ISO/IEC Guide 28 (2004). Conformity assessment -- Guidance on a third-party certification system for products.

4 IEEE 1262 (1995). Recommended Practice for Qualification of Photovoltaic (PV) Modules.

5 ISO/IEC Guide 40 (1983). General requirements for the acceptance of certification bodies.

6 IECQ. http://www.iecq.org/index.htm (Accessed 29 August 2019).

7 Varadi, P.F. (2014). *Sun Above the Horizon*. Singapore: PanStanford Publisher.

8 IECQ. https://www.iecee.org/about/what-it-is (Accessed 29 August 2019).

9 IECQ (2010). https://www.iec.ch/about/brochures/pdf/conformity_assessment/pv_certification.pdf (Accessed 29 August 2019).

10 IEC TS 62941 (2016). Terrestrial photovoltaic (PV) modules – Guideline for increased confidence in PV module design qualification and type approval.

11 Kelly, G., Spooner, T., Volberg, G. et al. (2014). Ensuring the reliability of PV systems through the selection of International Standards for the IECRE Conformity Assessment System. 40th IEEE PVSC in Colorado, USA (8–13 June 2014).

12 Kelly, G., Häring, A., Spooner, T. et al. (2015). Ensuring the reliability of PV systems through the formation of the IECRE Conformity Assessment System and the development of new International Standards. 42nd IEEE PVSC in Colorado, USA (14–19 June 2015).

13 IEC 62891 (2014). Indoor testing, characterization and evaluation of the efficiency of photovoltaic grid-connected inverters.

14 IEC 62109-1 (2010). Safety of power converters for use in photovoltaic power systems – Part 1: General requirements.

15 IEC 62109-2 (2011). Safety of power converters for use in photovoltaic power systems – Part 2: Particular requirements for inverters.

16 IEC 62093 (2005). Balance-of-system components for photovoltaic systems – Design qualification natural environments.

17 IEC 62852 (2014). Connectors for DC-application in photovoltaic systems – Safety requirements and tests.

18 IEC 60269-6 (2010). Low-voltage fuses – Part 6: Supplementary requirements for fuse-links for the protection of solar photovoltaic energy systems.

19 IEC 62548 (2016). Photovoltaic (PV) arrays – Design requirements.

20 IEC 62446 (2009). Grid connected photovoltaic systems – Minimum requirements for system documentation, commissioning tests and inspection.

21 IEC 62446–1 (2016). Photovoltaic (PV) systems – Requirements for testing, documentation and maintenance – Part 1: Grid connected systems – Documentation, commissioning tests and inspection.

22 IEC 62446–3 (2017). Photovoltaic (PV) systems – Requirements for testing, documentation and maintenance – Part 3: Photovoltaic modules and plants – Outdoor infrared thermography.

23 IEC 62446–2 (under development). Photovoltaic (PV) systems – Requirements for testing, documentation and maintenance – Part 2: Grid connected systems – Maintenance of PV systems.

24 IEC 61829 (2015). Crystalline silicon photovoltaic (PV) array – On-site measurement of I-V characteristics.

25 IEC 61724–1 (2017). Photovoltaic system performance – Part 1: Monitoring.

26 IEC 61724–2 (2016). Photovoltaic system performance – Part 2: Capacity evaluation method.

27 IEC 61724–3 (2016). Photovoltaic system performance – Part 3: Energy evaluation method.

28 IEC TS 62915 (2018). Photovoltaic (PV) modules – Type approval, design and safety qualification – Retesting.

29 IEC 62894:2014/AMD1 (2016). Photovoltaic inverters – Data sheet and name plate.

30 IEC 62817:2014/AMD1 (2017). Photovoltaic systems – Design qualification of solar trackers.

31 IEC 62738 (2018). Ground-mounted photovoltaic power plants – Design guidelines and recommendations.

32 IEC TS 63049 (2017). Terrestrial photovoltaic (PV) systems – Guidelines for effective quality assurance in PV systems installation, operation and maintenance.

33 Kelly, G., Häring, A., Spooner, T. et al. (2016). Coordination of International Standards with Implementation of the IECRE Conformity Assessment System to Provide Multiple Certification Offerings for PV Power Plants. 43rd IEEE PVSC in Colorado, USA (6–8 June 2016).

34 Kelly, G., Häring, A., Spooner, T. et al. (2017). Ensuring the Reliability of Photovoltaic Power Systems Using International Standards and the IECRE Conformity Assessment System. 44th IEEE PVSC in Washington, USA (25–30 June 2017).

35 Kelly, G., Häring, A., Hogan, S. et al. (2018). Defining the Value of IECRE Certifications for Providing Confidence in PV System Performance. WCPEC-7 in Hawaii, USA (10–15 June 2018).

36 IECRE. http://www.iecre.org/certificates/windenergy (Accessed 29 August 2019).

37 Kelly, G., Hogan, S., Ji, L. et al. (2019). IECRE Site Inspection Demonstration. NREL PVRW.

9

Predicting PV Module Service Life

What the PV industry would like is one set of accelerated stress tests that when successfully passed would ensure that a particular module type could survive in the field for 25 years. Implicit in that statement is the assumption that the test sequence is reasonably short. Most people asking for such a test sequence are probably thinking in terms of a few months of testing with an answer that spans a reasonable range of years (something like 25–35 years). This is not likely to happen because not all modules are designed and made the same way, nor are they deployed in the same geographic locations, so they are likely to degrade or fail via different mechanisms at different rates. McMahon et al. [1] came to this same conclusion as early as 2000.

Take a quick review of Chapter 2 and you realize that modules can and do degrade and/ or fail in many different ways. Each of these failure modes may be caused by a different set of environmental conditions. To predict the module's service life, we must predict it in relation to each and every one of the identified failure modes it is likely to experience. If we are going to use accelerated stress tests to predict service life in the field, we are going to have to understand the relationship between accelerated stresses and field exposure. Section 9.1 will address how to go about determining the acceleration factors for the different accelerated stress tests that are typically performed on PV modules.

Some of the failure modes listed in Chapter 2 may be caused by manufacturing issues. If every module were built exactly like the ones that are tested, this would not be an issue. In the real world, however, stuff happens. The design of the module itself, the equipment used in manufacturing and the quality control system used for production should minimize excursions of the process. Knowing the history of module manufacturing for a particular type of module will provide information as to how well controlled the process was. This will help to determine what manufacturing defects at what level are likely for a particular module type. Section 9.2 will discuss the impact of module design and control of the manufacturing process on module failure rates and how that impacts lifetime predictions.

All you have to do, is look at a world map of daily weather to realize how different the conditions are from place to place. Temperatures, humidity, wind speed, cloud cover, and precipitation vary based on the geographic location where a module is deployed. If any of the module's degradation or failure modes depend on these factors (and many do) then they will be stressed at different levels based on where in the world they are deployed.

Photovoltaic Module Reliability, First Edition. John H. Wohlgemuth.
© 2020 John Wiley & Sons Ltd. Published 2020 by John Wiley & Sons Ltd.

In addition, if a module is mounted in a way that restricts air flow and cooling, for example, on a rooftop or is building integrated, it will run at a higher temperature than open-rack mounted modules at the same geographic location. Section 9.3 discusses the impact of geographic location and mounting on module degradation and failure rates and how those impact lifetime predictions.

Sometimes, it is necessary to learn to crawl before you can walk. Instead of jumping directly from a qualification test that assesses infant mortality to a lifetime prediction test, it may be best to find an intermediate level of testing. This is the sort of work that has been ongoing in The International Photovoltaic Quality Assurance Task Force (PVQAT) and was discussed in Chapter 7. Can we get the PV community to agree on one set of extended stress tests that evaluate modules for wear out? Section 9.4 addresses development of extended stress testing of PV modules.

While this chapter began with a statement that finding one set of accelerated stress tests that could be used to predict a module's service live was unlikely, that doesn't mean that it is impossible to predict service life. The earlier parts of this chapter all explain why a single test sequence is not likely to be valid for all module types. However, it should be possible for a manufacturer to predict the service life for each of their module types. This will require knowledge of the manufacturing process and its controls, feedback from field deployment and extensive accelerated stress testing. Such an approach will likely take years, not months but the results will provide a baseline for evaluating future products of similar design. Section 9.5 will discuss how a PV-module manufacturer could set up a system to predict the lifetime of one of their products.

Finally, in order to have a program that predicts PV-module lifetime, we must define what is meant by the term. There are several requirements for a module to be considered as still functioning:

1) It must perform in a safe manner, particularly in terms of protecting personnel from high-voltage shock hazards and not cause fires.
2) It can still be deployed and produce electricity. This means that the frame and substrate or superstrate are intact so it can be mounted pointing at the sun.
3) It produces some predetermined amount of its original power. For this book, a value of 80% will be the arbitrary cutoff to be consistent with many module warranties.

With this definition of a module lifetime, what does it mean that we can predict a module service life? Certainly not all of the modules will still be working after this timeframe. This analysis has to be a statistical function that predicts that the overwhelming majority (something like 95%) of this type of module will survive at the 80% power level when deployed at a particular geographic location in a particular mounting system.

9.1   Determining Acceleration Factors

To use an accelerated stress test to estimate the lifetime of a product, the acceleration factor must be determined. Many degradation processes are activated by increasing temperature. If a single mechanism dominates over the temperature range of interest, the rate of degradation can be approximated by the Arrhenius equation:

$$R = C * \exp\left[-\frac{Ea}{kT}\right] \quad\quad\quad (9.1)$$

Where:

R = Reaction rate
C = A Constant
Ea = Activation Energy of the degradation reaction
k = Boltzmann's constant
T = Temperature in °K

To find the value of Ea, a series of measurements of the reaction rate are made at different temperatures. Then the values of ln(R) are plotted versus 1/T. This plot should result in a straight line with a slope equal to Ea. As long as the operating use temperature is known, the acceleration factor between accelerated test conditions and use conditions can be calculated.

One example where this relationship is used, is in the thermal degradation of diodes. As part of the specification for a commercial diode, the manufacturer provides a maximum junction temperature for continuous operation. As long as the diode stays below this temperature it should not suffer from thermal degradation. If it does operate at a higher temperature, the relationship in Eq. (9.1) can be used to predict its service life. For an appliance like a radio, this is fairly straightforward as the radio is either off so the diode is at room temperature or on so the diode is at the elevated temperature. The manufacturer can then estimate the service life based on hours of usage.

For the bypass diodes used in PV modules, the manufacturer's maximum junction temperature is used as the pass/fail for the bypass diode thermal test. Diodes must operate at a temperature below the rating during a period when it is heated to 75 °C while passing the short-circuit current of the module. This is similar to measuring the diode temperature in a radio when it is operating. So, this certainly provides a degree of safety for the diode. However, unlike the radio case, the bypass diodes have a more complicated life. The bypass diodes are safety devices and are in the "off" mode during normal operation of the module. They only turn on when the cells they are protecting are shadowed or there is a problem with that circuit (usually an open circuit). Many bypass diodes may only operate occasionally during their lifetime so thermal degradation is not likely to be an issue. On the other hand, some bypass diodes may pass current whenever the sun is shining because they are permanently shadowed (think of a chimney on a roof) or the circuit they protect has an open circuit. In such cases, the diode must survive or the string will lose much or all of its power and there may be the potential to start a fire. So, for the small fraction of diodes that are operating in the forward mode when the sun shines it is important that they do not degrade. The work by Shiradkar under PVQAT Task Group 4 presented in Chapter 7 [2] showed that our present bypass diode thermal test is perfectly adequate for diodes used in open-rack mount situations. However, it also showed that modules mounted with restricted cooling (rooftops and Building Integrated Photovoltaics (BIPV)) do run hotter than under the IEC 61215 bypass diode test conditions. This is being addressed in a new standard, IEC 63126: "Guidelines for Qualifying PV Modules, Components and Materials for Operation at High Temperatures," which will be discussed in Chapter 10.

Thermal degradation is not the only way the diode in a radio may fail. Someone could drop it or leave it out in the rain. Similarly, the bypass diodes in a PV module have other failure modes besides thermal degradation. These include failure of their electrical bonds to the rest of the module circuit, which is tested via thermal cycling, or exposure to a nearby lightning strike. Lightning failures of bypass diodes tend to be localized events dependent on the geographic location, the mounting system and the electric circuit they are installed into. So, we can address the bypass diode service life in terms of thermal degradation by defining the correct bypass diode thermal test, but there will be additional issues to address in some locations where lightning is prevalent.

Of course, solar cells are just very large diodes so thermal degradation of their junction also follows an Arrhenius type behavior. It is unlikely that the temperature of a PV module will ever reach a level where thermal degradation of the solar cell (particularly cry-Si cells) will occur. If there was some unusual condition that resulted in such a high temperature, a number of other components in the module would fail first, including solder bonds and many encapsulants. Ask someone who operated a PV test laboratory in the earlier days of PV what happened when their test chamber went out of control and heated to its upper limit? The results were not pretty and none of the exposed modules provide any output power afterwards.

Thermal degradation alone is not a major issue for PV modules. However, higher temperatures do accelerate many of the other degradation modes we will talk about. You can take it as a rule of thumb that everything else being equal, a module deployed in a hotter location will degrade more rapidly than one deployed in a cooler location.

It is also important to remember that in a solar module it is not usually the solar cell junction that fails. It is usually the electrical connections or one of the polymeric components that degrades resulting in lower efficiency or failure of the protective package.

Now, let's take a look at several failure mechanisms where models have been developed and acceleration factors have been calculated.

9.1.1 Thermal Cycling

As discussed in Chapter 7 Nick Bosco from NREL working with PVQAT Task Group 2 was able to develop a model that predicted the damage done by thermal cycles [3]. Applying this model to both accelerated stress testing and field exposure at various sites, he was able to equate the effects of both as shown in Figure 7.5. His work clearly indicates that the acceleration factors depend significantly on the geographic location in which the modules are deployed. While three thermal cycles (from −40 to +85 °C) are sufficient to provide the same stress as a one-year exposure in Sioux Falls, South Dakota (a fairly cool climate), more than 25 of the same cycles would be required to provide the same stress as a one-year exposure in Chennai, India (a hot climate with lots of intermittent clouds).

This shows that to predict service life, the weather is going to have to be considered (See Section 9.3). However, we could test for reliability against this failure mode by always testing for the worst case. For thermal cycling, we could use Chennai as the worst case and test all modules for 650 thermal cycles (−40 to +85 °C) for a 25-year minimum lifetime (26 cycles per year times 25 years). This long-term reliability approach will be discussed in Section 9.4.

9.1.2 Discoloration of the Encapsulant

As also discussed in Chapter 7, PVQAT Task Group 5 conducted a set of experiments using accelerated UV testing to duplicate the change in optical transmittance for the historical ethylene vinyl acetate (EVA) formulation (STR-A9918) that discolored in the field. The change in solar-weighted transmission was a linear function of radiant exposure (See Figure 7.9) and was dependent on the temperature with an activation energy of ~60 kJ/mol as shown in Figure 7.10. This result is consistent with the degradation in short-circuit current (~0.5%/year) measured in modules exposed in the field for more than 25 years in California.

If anyone was going to produce cry-Si PV modules using the old A9918 formulation of EVA, we can now calculate the expected reduction in solar-weighted transmission and therefore the reduction in short- circuit current based on the type of mounting system used and the weather in the geographic area where deployed. That, however, is not likely to happen as non-discoloring EVA formulations are available so why would anyone knowingly use one that will degrade? In this case, understanding the degradation mechanism resulted in development and use of encapsulant technologies that eliminated the problem.

9.1.3 PET Hydrolysis

Polyethylene terephthalate (PET) is a commonly used backsheet in PV modules because of its relatively low cost and good dielectric properties. However, there have been reports of PET backsheets degrading during prolonged damp heat testing. Some in the PV industry, began to question whether PET or polyester was a good choice for a backsheet even though modules with PET layers have survived in the field for more than 30 years. Michael Kempe of NREL set out to model the PET hydrolysis reaction to determine whether PET degradation after 2000 hours at 85 °C and 85% Relative Humidity should be considered an issue for PET survival in the field.

Polyester is susceptible to chain scission via a hydrolysis reaction that leads to degradation of the mechanical properties. The kinetic rate of hydrolysis was examined by McMahon [4] and he developed an equation for the reduction in bonds caused by the hydrolysis. In separate work that yielded similar degradation rates, Pickett [5, 6] found that the reaction was second order with respect to the relative humidity of the water vapor present. Kempe used Pickett's equation and McMahon's data to calculate an activation energy of 129 kJ/mol for the hydrolysis reaction. This compares well with Pickett's values of 125–151 kJ/mol that he found for four different grades of PET.

McMahon determined that PET became brittle (determined by a one-third initial tensile strength loss) and "failed" when $\log10(C/C-x) = ~0.0024$, or when ~0.55% of the ester bonds have been broken by hydrolysis. By using the 0.55% failure criterion along with Pickett's equation, one can estimate the service life of PET with respect to embrittlement via hydrolysis. Kempe [7] carried out an analysis using the module (cell) temperature along with relative humidity calculated as the average of the outside surface of the module and that in the back-EVA sheet, to model the exposure of a PET backsheet on the backside of a module. The results are given in Table 9.1, where the degradation caused by 1000 hours of 85 °C/85% relative humidity (RH) exposure is compared to how many years it would take

Table 9.1 Modeling of polyethylene terephthalate (PET) hydrolysis for open rack mounted modules.

Geographic location	Years Outdoor exposure equivalent to 1000 hours of Exposure at 85 °C & 85% RH
Denver, Colorado	6500
Munich, Germany	5100
Albuquerque, New Mexico	4400
Riyadh, Saudi Arabia	4000
Phoenix, Arizona	1700
Miami, Florida	530
Bangkok, Thailand	320

Source: from Ref [7].

to suffer the same level of degradation for modules in an open-rack mounted configuration at seven different geographic locations around the world. It is obvious from these results that the 1000 hours of damp heat testing used as part of the qualification test sequence is already overkill for a 25- or 30-year service life prediction for polyester hydrolysis anywhere in the world.

Kempe's modeling also indicated that PET should begin to degrade due to hydrolysis after about 2000 hours of exposure at 85/85. So, his modeling is consistent with the accelerated test results. We do not expect PET hydrolysis to be an issue for module lifetime when mounted in a normal fashion anywhere on earth. This does not mean that PET or any other backsheets are for sure going to last 25 or 30 years in the field. There are other combinations of stresses, like UV and temperature that can cause backsheet degradation.

So, we have developed models for the acceleration factors involved in three of the degradation/failure modes. In Chapter 2, we identified many more degradation/failure modes. To fully predict the lifetime of a particular module type, at a particular location, we would have to know the exact tests to perform for every degradation/failure mode, as well as knowing the acceleration factors for each of those tests. This is beyond the scope of what we can do today though there are other degradation/failure modes on the list that are being studied with the eventual hope of identifying the required accelerated test and its acceleration factor.

Any time you are dealing with accelerated stress tests, the question of the uncertainty of the results should be addressed. This is one of the most critical factors involved in defining the duration of your testing, especially for those looking to find the tests with the shortest duration. A general statement in this regard is that the shorter the test (that with the highest acceleration factor) the higher the uncertainty. This incorporates two factors. The first being the fact that there is always a degree of uncertainty in the determination of the acceleration factor and that uncertainty is always greater for higher acceleration rates. The second reason is, that with a highly accelerated stress test, you can never quite be sure that the failure mode observed under the accelerated stress test is the same one that will occur in the field and be the lifetime limiter. So, at higher acceleration rates you may actually be measuring the rate of the wrong failure/degradation mode. This is especially true for complex failure/degradation modes that depend on multiple stresses like temperature, humidity, UV irradiance or applied voltage.

Michael Kempe of NREL has presented several talks on evaluating the uncertainty during accelerated stress testing [8, 9]. Some of his observations include:

- The only time you can reach uncertainty levels of ± 1 or 2 years is when you have an extremely good understanding of the degradation kinetics. This is unusual and is often restricted to a limited situation – one product mounted by one method at a particular geographic location.
- For exposed module materials like backsheets, a long generic test (~2 years) may yield lifetime estimates with reasonable uncertainties on the order of ± 8 years.
- Accelerated stress tests can either be targeted at specific materials and failure modes in certain geographic locations, or of necessity, they must be of a very long duration to achieve reasonable uncertainties.

9.2 Impact of Design and Manufacturing on Failure or Degradation Rates for PV Modules

The design of a PV-module type and method of manufacturing including the quality management system (QMS) are important factors in predicting the lifetime. Many of the impacts of design on the lifetime are taken for granted today. For example, the use of multiple bus bars (often five or even six per cry-Si cell) with multiple solder bonds per ribbon provide a great degree of redundancy that minimizes the degradation from a few failed solder bonds or even cracks in the solar cells. Similarly, with the manufacturing process, we expect control of the process to be sufficient for the production modules to continue to pass the qualification (IEC 61215) and safety (IEC 61730) tests. But it is really only the module manufacturer who knows when design and process changes were made or what out of control conditions happen on the production line and what dangers those out of control conditions represent.

We can probably all understand why it is important to know when design changes were made in a module type. These are often not obvious from just looking at a module. We would certainly notice if the cells used in a module switched from four to five bus bars. However, we are not likely to identify a switch from EVA with UV stabilizer to EVA without UV stabilizer. But it is critical that the module manufacturer knows exactly when this change was made so the result of both accelerated stress testing and field deployment can be categorized for the two different EVA's. This way, the manufacturer can determine if the change has a negative or positive impact on module performance and reliability. Such an impact may take many years to occur in the field, requiring comparison of accelerated test results with field deployment results over many years.

There are many reasons why a module manufacturer makes a process change. Often, it is to save money or to increase production throughput, but it may also be to improve performance (either initial or long term via reducing degradation). These changes should, as a minimum, be qualified through IEC 61215 and IEC 61730 before being implemented. However, if the change is directly related to the module's ability to operate reliably for its design life, it should be tested in one of the extended test sequences for wear-out (See Section 9.4). In addition, the fielded modules should be followed for years to see whether the change had an effect on lifetime or degradation rate.

The third area of concern is process control and manufacturing defects. In looking at the list of defects in Table 2.1, a number of them could be due either to manufacturing defects or to out of control processes. Manufacturing defects may include:

- Broken or cracked cells
- Bad solder joints especially those on termination ribbons as shown in Figure 2.12
- Poor junction box adhesion as shown in Figure 2.17
- Poor frame attachment

Out of control processing defects may include:

- Delamination if the primer hasn't been adequately activated or the encapsulant properly cured
- Bad solder joints on the ribbons as shown in Figure 2.13
- Broken or cracked cells
- Potential Induced Degradation (PID) if the cell anti-reflective (AR) coating process is out of control.

Only the module manufacturer knows how hard the processes are to control and to perform correctly, how well trained the operators are and finally what happens when control is lost. Are the "bad" products destroyed, reworked, shipped as seconds or possibly shipped as is.

To summarize this section, there is information about the way the products are built that only the manufacturer knows. This information is likely to be a critical part of predicting the lifetime of these products.

9.3 Impact of Location and Type of Mounting on Failure or Degradation Rates for PV Modules

Many of the examples of module degradation depend on the environmental conditions that the module experiences like temperature, humidity, irradiance levels, wind speeds, etc. It follows that the weather at the deployment site will be important in predicting service life. The first step is to identify the weather conditions at the geographic location of interest. These can be obtained from a local weather station or from satellite data. In either case, it is important that the actual instantaneous data be used, not averages. When averaging, the highs and lows are removed and everything looks mediocre, which tends not to be as stressful as the "actual" weather. Nick Bosco's work on thermal cycling discussed in Chapter 7 [3] showed that to accurately count temperature reversals, the temperature data had to be collected in minutes not hours. So, use of one- or five-minute intervals for weather data will probably yield the best results.

Module Temperature: Usually, the weather data includes air temperature but not module temperature which is really the value of interest. So, we have to calculate module temperature from the weather data that we have. There are two recognized models for doing this.

1) The King Model [10] uses Eq. (9.2).

$$T_{mod} = T_{amb} + I_{rr} * exp\{-a - b * W\} + \Delta T * I_{rr}/1000 \qquad (9.2)$$

Where:

T_{mod} is the module temperature to be calculated.
T_{amb} is the ambient temperature of the air from the weather data.
I_{rr} is the plane of array irradiance on the module calculated from the horizontal irradiance usually available as part of the weather data.
W is the wind speed from the weather data.
a, b, and ΔT are constants determined for each different module construction (for example glass/encapsulant/polymeric backsheet) as it is mounted (for example open-rack, roof or fully insulated). The values for these constants are given in the King reference.

2) The Faiman Model [11] uses Eq. (9.3).

$$T_{mod} = T_{amb} + I_{rr}/\left(u_0 + u_1 * W\right) \qquad (9.3)$$

Where:

T_{mod}, T_{amb}, I_{rr}, and W have the same meaning as for the King Model.
u_0 is a constant defining the impact of irradiance on the module temperature.
u_1 is a constant defining the impact of wind speed on the module temperature.

Once again, the values of u_0 and u_1 depend on the module construction and the mounting. Values for these can be found in the literature or measured using the procedures given in IEC 61853-2 [12].

Sarah Kurtz of NREL led a PV-industry effort to better understand PV module operating temperatures in order to investigate the effects of long-term, cumulative exposure to these temperatures. This work assumed Arrhenius type behavior with a range of reasonable activation energies [13, 14]. The study used typical weather data for eight locations around the world, modeled the module temperatures for six configurations of construction and mounting, and from these, calculated effective degradation temperatures. Module temperatures were calculated using the King Model [10].

Graphs showing the hourly temperature profiles at the eight geographic sites for each of the six configurations were published in the papers. The graph for a glass/cell/polymeric backsheet construction mounted in an open-rack configuration is shown in Figure 9.1.

Some conclusions that can be drawn from this analysis are:

• The open-rack temperature profiles were very similar and nearly independent of the module construction.

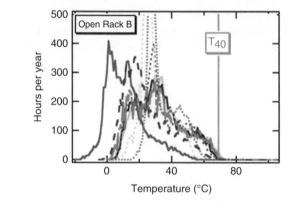

Figure 9.1 Hours per year module will operate at given temperature for glass/cell/polymeric backsheet in open rack configuration [13]. *Source:* reprinted from PVSC paper with permission of Sarah Kurtz of University of California at Merced.

- The maximum temperatures are similar to that predicted from a simple calculation of the operating temperature with an ambient temperature of 40 °C, an irradiance of one sun and no wind.
- In open-rack configuration the module temperature almost never exceeds 70 °C.
- Module temperature distributions are complex functions.

How can we predict the impact of operating with such a complex temperature profile? If the degradation follows the Arrhenius relation (as many thermal degradations do), it is possible to define an equivalent temperature that represents the degradation that would have taken place if the module had been aged for the same time period at a constant or equivalent temperature. The equivalent temperature can be calculated using Eq. (9.4) [13–15].

$$\exp\left\{-E_a/kT_{eq}\right\} = \frac{1}{t_1 - t_2} \int_{t_1}^{t_2} \exp\left\{-E_a/kT_m\{t\}\right\} dt \tag{9.4}$$

Where:

T_{eq} is the Equivalent temperature
E_a is activation energy of the degradation process
t is time
$T_m(t)$ is the time dependent temperature of the module

The equivalent temperature can tell us about the relative degradation rates at different geographic locations in different mounting configurations. However, as Eq. (9.4) shows, to calculate the equivalent temperature we need the activation energy of the degradation mode. Usually, we do not have this value. Based on work by Dixon [16], Kurtz and team selected three activation energies (0.6, 1.1, and 2 eV) over the range observed in Dixon's work. Using these three activation energies, they calculated the equivalent temperatures

Table 9.2 Equivalent temperature (°C) calculated for two module constructions, mounted in open rack at four geographic locations.

Locations	Glass-cell-glass			Glass-cell-polymer			Aver ambient
E_a	0.6 eV	1.1 eV	2.0 eV	0.6 eV	1.1 eV	2.0 eV	
Riyadh	42.2	48.4	55.1	40.3	45.9	52.3	26.2
Phoenix	40.6	47.1	53.9	38.7	44.6	51.0	23.8
Miami	33.5	37.1	41.9	32.1	35.0	39.3	24.3
Munich	19.4	26.7	35.4	18.2	24.8	33.0	8.0

Source: data from Ref [14].

for six module construction/mounting configuration combinations using weather data from all of the eight geographic locations [14]. Let's take a look at a few particular comparisons to get a better understanding of what this analysis can tell us.

Table 9.2 compares modules with glass/cell/glass construction to those with glass/cell/polymeric backsheet construction mounted in an open-rack configuration. To simplify, only four geographic locations are shown, Riyadh the hottest, Phoenix the hottest in the US, Miami a warm, humid location and Munich the coldest of the eight locations modeled. A few of the observations from this table:

- The equivalent temperature for glass/glass modules runs a few degrees Centigrade hotter than for glass/polymeric backsheet modules.
- There is quite a difference between the equivalent temperatures in the four locations with Riyadh having appreciably higher equivalent temperatures than Munich and even Miami though Riyadh and Miami have average temperatures that are not that different.
- The differences in equivalent temperature, due to the changes in activation energy, are at least as large as those due to change in geographic location. This is one of the reasons why it is so hard to select parameters for accelerated stress tests when you don't know the exact failure modes.

As a second example, let's look at the same module construction (glass/cell/glass) mounted in two different configurations, open rack and roof mount. These results are shown in Table 9.3 for the same four locations. Some observations from this table:

- Roof-mounted modules have an equivalent temperature that is 10–15 °C higher in the hotter climates (Riyadh and Phoenix) with less increase in cooler climates (Munich).
- Roof-top operation in Munich has a lower equivalent temperature than open-rack operation in Riyadh or Phoenix. So, roof-top operation in cooler climates should be no more stressful than open-rack operation in hotter climates.
- The change in actual activation energy has an even greater impact on the equivalent temperature on the roof top than it did for open rack.

This analysis should help us to better understand the impact of local temperature on the reliability/durability of PV modules.

Table 9.3 Equivalent temperature (°C) calculated for two mounting configurations (open rack and roof mount) at four geographic locations for glass–glass construction modules.

Mount	Open rack			Roof mount		
Location	0.6 eV	1.1 eV	2.0 eV	0.6 eV	1.1 eV	2.0 eV
Riyadh	42.2	48.4	55.1	52.0	61.0	69.5
Phoenix	40.6	47.1	53.9	50.6	59.9	68.2
Miami	33.5	37.1	41.9	40.4	46.7	53.7
Munich	19.4	26.7	35.4	26.3	37.2	48.2

Source: data from Ref [14].

Module Humidity: Once again, the weather data provides the humidity level in the local environment but not within the module. When something like a module is heated, it tends to drive out moisture and dry out. The inside of PV modules will typically have lower levels of humidity than the ambient air during the warmest part of the day when the sun is heating the modules above the local environmental air temperature. We have to determine the humidity levels inside the module rather than use the values directly from the weather data. This is more complicated than the calculation for temperature. The whole module is pretty much at the same temperature within a degree or two, but different parts of the module are typically at different humidity levels. Many polymeric backsheets are porous to humidity so the back of the module reaches humidity equilibrium quickly. On the other hand, the area between the cells and front glass in a cry-Si module has restricted access and so the humidity has to diffuse in slowly from the sides. Different parts of the module and different module constructions will require different models for humidity ingress.

Michael Kempe of NREL has published a number of papers describing the process of modeling moisture ingress into PV modules and has provided results of such modeling for a variety of module constructions in various environments [17, 18]. The start of any such modeling process is to determine the properties of the materials of which the module is constructed, particularly, the diffusivity and solubility of the packaging materials. This usually means as a minimum knowing the values for the encapsulant, backsheet and frontsheet. Since a majority of modules use EVA as the encapsulant, much of Kempe's work has used the values for EVA, though the general results of the modeling would hold for many of the other encapsulants in use today. For frontsheets and backsheets, we have to determine whether they are polymeric and therefore permeable (breathable) or made of glass or have a vapor barrier like Al foil and so are impermeable to moisture ingress.

One of the first constructions that Kempe evaluated, was the standard cry-Si module with a glass superstrate, EVA encapsulant and polymeric backsheet composed of Tedlar, PET and EVA. His modeling showed that the moisture penetrated the entire rear of the module within 24 hours and that after the first day, there were daily oscillations around an equilibrium value. The humidity is entering at night and is released during the day.

There is no build-up of humidity over time – it continues to oscillate around an equilibrium value determined by the local weather conditions.

The situation is different if we look at the constricted space between cell and glass or for a glass–glass module itself. In this case, the moisture must diffuse in from the sides. This can take most of a year in Bangkok, Thailand as was shown in Figure 6.5 [19]. In some drier climates, the level of moisture actually continues to increase over a number of years, but it eventually saturates at a level consistent with the average in the local climate.

On the other hand, when we perform an extended humidity test, the humidity level within the module continues to increase until it reaches a level consistent with the level in the chamber (85% RH at 85 °C) as was shown in Figure 6.4 [19]. This results in a water content of $0.0055\,g/cm^3$ which is four times higher than the highest level observed within the module in Bangkok, Thailand; one of the most humid places on earth. So, care must be taken when trying to equate damp heat test results to in-field performance.

9.4 Extended Stress Testing of PV Modules

Chapter 7 talked about the need for extended stress testing of PV modules, but indicated that, in the past, the tests from IEC 61215 were just extended without field evidence and/or modeling to validate that the selected additional stress levels were justified. This section will first look at two specific cases where sets of extended stress tests were developed based on engineering analysis of real-world data. Then, it will discuss the efforts underway in IEC TC82 to develop a consensus standard for extended stress testing of PV modules.

Qualification Plus: In 2013, the Department of Energy challenged NREL to review the status of PV module accelerated stress testing and to fill an immediate need by developing a well-defined set of accelerated stress tests that correlate with the field performance of PV modules [20]. In establishing Qualification Plus, the following were considered essential [21]:

- All included tests must be related to observed field failures of cry-Si module types that have successfully passed the qualification test.
- All tests must either be international standards or under development to become international standards.
- Inclusion of random sampling to eliminate cherry picking of test articles.
- Testing on a continuing basis rather than a one-time event.

Every effort was made to avoid extended tests unless there was a technical basis to demonstrate relevance to field performance.

The tests in Qualification Plus include module-level tests like those in the qualification test sequences (IEC 61215) as well as material and component-level tests like those in the module safety standard (IEC 61730-1 and IEC 61730-2). Ultimately, the safety tests should include longer-term material tests like those proposed for Qualification Plus but as you will see in Chapter 10, this is still in process. The emphasis is on testing the functionality of each material to show that it can perform as it is designed to perform within the module both before and after exposure to the relevant stresses. The material tests in Qualification Plus include:

UV Exposure of Encapsulants: Discoloration of module encapsulation in fielded modules was discussed in Chapter 2 as an observed field degradation mode and, again in Chapter 7, as part of the PVQAT work to develop a test to evaluate for it. Qualification Plus was written before the PVQAT Task Group 5 work was completed, but the test procedures are similar. The Qualification Plus test conditions for UV testing of encapsulants are:

- Irradiance of $56 \pm 5 \, \text{W/m}^2$ between 300 and 400 nm, or $0.55 \pm 0.05 \, \text{W/m}^2/\text{nm}$ at 340 nm
- Chamber controlled to ambient air temperature of $70 \pm 5\,°\text{C}$.
- Chamber humidity of $50 \pm 10\%$ at air temperature.
- Corresponding Black Panel (uninsulated) Temperature of $90 \pm 5\,°\text{C}$
- Duration of exposure: ~4000 hours to achieve at least $224 \, \text{kWh/m}^2$ total UV (300–400 nm).

The pass/fail is based on <2% decrease in solar-weighted photon transmittance after the exposure.

UV Exposure of Polymeric Materials: Long-term exposure of polymers to sunlight can cause some of these materials to degrade. Since these polymers are often part of the module electrical insulation (backsheet, junction boxes, connectors, etc.) their failure can result in potential safety hazards. The Qualification Plus tests apply a longer UV exposure at a moderately low temperature to these polymers. The components and their test conditions are:

- Cables and connectors use the same exposure as encapsulants.
- Backsheets use an irradiance of $81 \pm 8 \, \text{W/m}^2$ between 300 and 400 nm, or $0.8 \pm 0.08 \, \text{W/m}^2/\text{nm}$ at 340 nm with similar temperatures and humidity level as encapsulants. The duration of the exposure is ~4000 hours to achieve at least $320 \, \text{kWh/m}^2$ total UV (300–400 nm). Since backsheets can be exposed from both the front and back sides, Qualification Plus requires exposure from both sides. Front side exposure is through the intended glass/encapsulant combination while the back side exposure is directly onto the rear side of the backsheet.
- Junction boxes use the test conditions in the IEC junction box standard (IEC 62790) which includes water spray as part of the cycle.

IEC 61215 contains one component test, the bypass diode thermal test. At the time of writing Qualification Plus, there was concern that the test was not severe enough because significant diode field failures had been reported [22] in roof-mounted systems. To address this, Qualification Plus increased the duration of the Bypass Diode Thermal Test from 1 to 96 hours under the same stress conditions. As was reported in Chapter 7, the work of PVQAT Task Group 4 has now shown that it is not the duration of the test that is the problem but rather that roof-top arrays operate at higher temperatures than the test temperature of the Bypass Diode Thermal Test. So, a modified Bypass Diode Thermal Test is now being proposed for higher temperature deployment (See Chapter 10).

Module Testing: Qualification Plus increased stresses that modules are expected to experience in the field in order to identify some of the most commonly observed wear-out field failures. It also added several new tests based on failure modes observed in the field but that had not yet been included in IEC 61215 or IEC 61730. Figure 9.2 shows a diagram of the module tests in Qualification Plus.

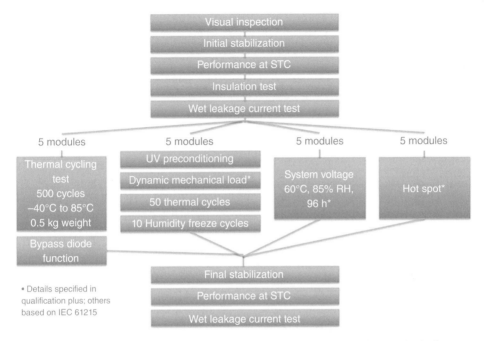

Figure 9.2 Module level tests for qualification plus [20]. *Source:* reprinted from author's diagram used in NREL Report.

The changes/additions to the test sequence in Qualification Plus are:

- The number of Thermal Cycles has been increased from 200 to 500. This was done because field data indicated that modules that survive 200 thermal cycles from the qualification test have suffered broken interconnects and degraded solder bonds in the field. [23]. This is also consistent with the results of PVQAT Task group 2 work discussed in Chapter 7 that led to publication of IEC 62892 [24].

- The Dynamic (Cyclic) Mechanical Loading (DML) Test (IEC 62782) [25] was added to the test sequence after the UV exposure and before the 50 thermal cycles. Dynamic mechanical loading, followed by thermal cycling, has been used as a stress test for cell breakage and other potential failures caused by bending and vibrating during manufacture, handling, transportation, installation, and operation [26]. This change has now been proposed for the next edition of IEC 61215 as will be discussed in Chapter 10.

- A PID test has been included in Qualification Plus. As indicated in Chapter 2, PID is a fairly recent issue and so was not included in the 2016 edition of IEC 61215. However, there is an IEC Technical Specification for PID of cry-Si modules (IEC TS 62804-1) [27]. Qualification Plus specified the chamber method from IEC TS 62804-1. The addition of a PID test to IEC 61215 has been proposed for the next edition as will be discussed in the Chapter 10, but the proposed test parameters have been modified to make the test more stressful.

- The Hot Spot Test in the 2006 edition of IEC 61215 had problems. Qualification Plus called for the use of ASTM E2481-06 [28] which, in reality, was a draft of the changes that were eventually made in the 2016 edition of IEC 61215.

IEC 61215 allows for the use of engineering samples to validate the design. Qualification Plus, however, is designed for manufactured product, not pre-production. Therefore, the samples used for the testing must be randomly selected from a production line that is actively shipping products. Because it is production sampling, a larger statistical sample was desired. Qualification Plus requires testing of five modules through each of the test legs instead of the two used in IEC 61215. Finally, while IEC 61215 allows retesting of one test leg, in case of a failure in that test leg, Qualification Plus requires a full retest if one module fails one test. A one-in-five or 20% failure rate in the field would not be acceptable to most customers so why should it be allowed in a test of product quality?

Finally, under Qualification Plus, manufacturers are required to meet the requirements of IEC TS 62941 [29] and have their quality assurance system audited to this technical specification by a Certification Body approved by ANSI-ASQ National Accreditation Board (ANAB) to conduct ISO 9001 Quality System Audits or a Certification Body approved by a similar organization under the International Accreditation Forum (IAF).

A number of module manufacturers had products certified under the Qualification Plus system. However, it was really designed as an interim system so it has now been supplanted by more recent schemes.

CSA C450 – PV Module Durability Testing Protocol and Methodology [30]: In 2016, CSA Group (Canadian Standards Association) organized a committee of PV experts with representatives from National Laboratories, module manufacturers, test laboratories and banks to develop an extended stress test sequence. Their goal was to publish a publically-available specification that reflects the best practices in the industry for testing modules for Quality Assurance, reliability and durability. Individual tests were all supposed to be taken from existing standards addressing qualification, safety and long-term durability, mostly from IEC, but also drawing on Qualification Plus.

The CSA document draws heavily on IEC 61215, basically copying its requirements for Test sample selection, Module markings and documentation, Pass/fail criteria and the definition of major visual defects. One of the major departures from IEC 61215 is the requirement to test four modules in each of the test legs rather than just two. Another is the requirement to take electroluminescence (EL) images before and after each test to better evaluate what is changing within the module.

C450 specifies a number of characterization tests before each sequence, as interim tests and as final tests. These include:

Initial and Final Characterization: Visual Inspection (MQT 01 from IEC 61215), EL imaging, Stabilization (MQT 19), Standard Test Conditions (STC) Power Measurements (MQT 06), Insulation Test (MQT 03) and Wet Leakage Current Test (MQT 15).

Interim Characterization: Visual Inspection (MQT 01), EL imaging, STC Power Measurements (MQT 06), Insulation Test (MQT 03) and Wet Leakage Current Test (MQT 15).

C450 has five parallel legs of tests that are described below:

Leg 1: Thermal Cycling – This leg calls for three successive sets of 200 thermal cycles from −40 to +85 °C with current flow (MQT 11). After the first two sets of 200 cycles, the modules are characterized per set of interim characterization procedures. After the final set (for a total of 600 TC's), the modules are characterized using the set of final characterization procedures.

Leg 2: Mechanical Durability – This leg is similar to the UV/TC/HF leg in IEC 61215 but the UV is moved to a different leg and the Dynamic (Cyclic) Mechanical Loading Test is performed first. So, the sequence starts with 1000 cycles of IEC 62782 [25] using a force of 1500 Pa followed by an interim characterization set. Then the modules are subjected to 50 Thermal Cycles from −40 to +85 °C (MQT 11) followed by another interim characterization set. Finally, the modules are subjected to 10 humidity-freeze cycles from −40 to +85 °C at 85% RH (MQT 12) followed by the final characterization set.

Leg 3: Humidity Freeze UV Leg – This leg is taken from Sequence B in IEC 61730-2 but, in this case, the modules must meet the performance pass/fail after the stresses. The first step in this leg is a 200-hour exposure to damp heat at 85 °C/85% RH (MQT 13). The next step is a front-side UV exposure of 60 kWh/m^2 at 65 °C (MQT 10). The third step consists of 10 humidity-freeze cycles from −40 to +85 °C at 85% RH (MQT 12). These first three steps are followed by an interim characterization. Following the interim characterization, the back sides of the modules are exposure to a UV exposure of 60 kWh/m^2 at 65 °C (MQT 10). This is followed by another 10 humidity-freeze cycles from −40 to +85 °C at 85% RH (MQT 12) before the final characterization.

Leg 4: Damp Heat – This leg contains two exposures to the 1000-hour damp heat test at 85 °C/85% RH (MQT 13). The first 1000-hour exposure is followed by an interim characterization. The second 1000-hour exposure is followed by the final characterization.

Leg 5: PID – This leg consists of PID testing at 85 °C/85% RH at both plus and minus systems voltage for 192 hours using the methodology in IEC 62804-1 [27] followed by the final characterization.

At least one module manufacturer has reported qualifying their modules to C450.

IEC TS 63209 – Extended-stress testing of PV modules for risk analysis: In 2018, IEC TC82 approved initiation of this new project. According to the scope of this document, "the Technical Specification describes a data collection methodology to identify photovoltaic module strengths and weaknesses by applying stresses and characterizing changes caused by those stresses" [31]. So, there is no set of pass/fail criteria, just various sets of accelerated stress tests taken from published standards such as IEC 61215, IEC 61730, and any others deemed useful by the project team. The initial draft document makes it clear that this Technical Specification (TS) is not designed to:

- Replace IEC 61215 as a pass-fail test procedure for market entry, but rather to provide standardized methods of testing for those wishing to have more information than is provided by IEC 61215.
- Identify all module failure modes. It is not possible to do a large number of accelerated stress tests without causing some changes that are irrelevant to field performance. These may mask some changes that are important.

- Be used for service-life predictions because different products exhibit different failure mechanisms in different climates. Quantitative lifetime predictions need to be product, geographic location, mode, and/or mechanism specific. Service life predictions will require more complex sets of tests.

The TS is intended to provide data on potential failure modes for a particular module type. The data are designed to be analyzed to uncover probable causes of module failure and what situations are most likely to lead to those types of failure. It is intended to be used to make comparisons between products.

This project is just in its infancy so most of the details have yet to be worked out. It envisions several rounds of testing, starting with use of the standard qualification tests (to be identified by the project committee, using existing IEC standards such as IEC 61215, IEC 61730, and related tests), which will be applied and the results documented, including any failures. It isn't clear whether this means just documenting the results of the qualification tests as a first round, or if this will be some more elaborate set of tests based on the qualification tests. For example, you could repeat each of the legs in IEC 61215 multiple times until the modules began to degrade.

After this initial set of tests, the draft document indicates there may be multiple additional rounds of data collection using stresses or combination of stresses selected by the project committee based on common acceptance of those that will provide useful data. Work on the IEC extended stress test protocol is just beginning.

9.5 Setting Up a True Service Life Prediction Program

As was indicated at the beginning of this chapter, it is unlikely that a single set of accelerated stress tests will be able to predict the service life for all module types deployed in all locations. Hopefully, as the chapter unfolded, you now understand why this statement is true. However, all is not lost, as there are still ways to move forward toward lifetime predictions.

The first approach is the simplest and most straightforward, but the least useful. You deploy the modules, monitor their performance and wait, and wait, and wait until they finally degrade and/or fail. Of course, this should take 25 or 30 or more years and while giving you an accurate measure of the lifetime, does not allow you to do any predicting at least in respect to this first array deployed. One would assume that after 25 or 30 years of waiting, the technology has changed enough that the results are not particularly useful for deployment of new modules. So, we really need something that accelerates the process.

If we are interested in estimating the lifetime of a particular module type in a defined geographic area, we need to start by

- Assessing the QMS under which they were built.
- Reviewing the results of the qualification tests or performing them if this is a new product. If any extended stress testing has been done on these modules, those results should also be reviewed.
- Reviewing any field deployment results available for the climate of interest or deploy some modules in the climate of interest.

- Analyze the failure and degradation results available on this module type, both from field data and from various accelerated stress tests. Consider other modules types that have similar construction but may have more failure or degradation data available for comparison.
- Select one or several of the most likely failure/degradation modes. Perform the testing outlined in Section 9.1 to determine the acceleration factor for those most likely failure/degradation modes.
- Using these acceleration factors, use the accelerated stress test results to predict lifetime in the climate of the geographic region of interest. This will yield lifetime predictions for each of the selected most likely failure/degradation modes. The failure/degradation mode that yields the shortest lifetime prediction is the one most likely to result in end-of-life for this product.

This work depends on whether your analysis of the most likely failure/degradation modes is correct. Therefore, just completing the process to get an initial answer is probably not enough. Continued monitoring of field results coupled with extended stress testing guided by the field results will help continue refining the results. You need to identify any failure modes that are important, but that you did not suspect were in the beginning, and ultimately use this knowledge to improve the accuracy of the lifetime predictions.

References

1 McMahon, T.J., Jorgensen, G.J., Hulstrom, R.L. et al. (2000) Module 30 year life: what does it mean and is it predictable/achievable? *Proceedings of the 2000 National Center for PV Annual Meeting.* https://www.researchgate.net/publication/241896489_Module_30_year_life_What_does_it_mean_and_is_it_predictable-achievab.

2 Shiradkar, N., Gade, V., Schneller, E.J., and Sundaram, K.B. (2015). Revising the Bypass Diode Thermal Test in IEC 61215 Standard to Accommodate Effects of Climatic Conditions and Module Mounting Configurations. 42nd IEEE PVSC in New Orleans, USA (14–19 June 2015).

3 Bosco, N., Silverman, T., and Kurtz, S. (2016). Climate specific thermomechanical fatigue of flat plate photovoltaic module solder joints. *Microelectronics Reliability* http://dx.doi.org/10.1016/j.microrel.2016.03.024.

4 McMahon, W., Birdsall, H.A., Johnson, G.R., and Camilli, C.T. (1959). Degradation studies of polyethylene terephthalate. *Journal of Chemical & Engineering Data* 4: 57–79.

5 Pickett, J., Davis, M., and Zhou, J. (2011). Polycarbonate stability in solar energy applications. ATLAS/NIST PV Material Durability Workshop in Maryland, USA (27–28 October 2011).

6 Pickett, J.E. and Coyle, D.J. (2013). Hydrolysis kinetics of condensation polymers under humidity aging conditions. *Polymer Degradation and Stability* 98: 1311–1320.

7 Kempe, M.D. and Wohlgemuth, J.H. (2013). Evaluation of Temperature and Humidity on PV Module Component Degradation. 39th IEEE PVSC in Florida, USA (16–21 June 2013).

8 Kempe, M.D. (2014). Evaluation of the Uncertainty in Accelerated Stress Testing. 40th IEEE PVSC in Colorado, USA (8–13 June 2014).

9 Kempe, M. (2014). Using Uncertainty Analysis to Guide the development of Accelerated Stress Tests. NREL PVMRW.

10 King, D.L., Boyson, W.E., and Kratochvil, J.A. (2004). *Photovoltaic Array Performance Model*. Sandia National Laboratories, SAND2004-3535.

11 Faiman, D. (2008). Assessing the outdoor operating temperature of photovoltaic modules. *Progress in Photovoltaics* 16 (4): 307–315. https://doi.org/10.1002/pip.813.

12 IEC 61853-2 (2016). Photovoltaic (PV) module performance testing and energy rating – Part 2: Spectral responsivity, incidence angle and module operating temperature measurements.

13 Kurtz, S., Whitfield, K., Miller, D. et al. (2009). Evaluation of Cumulative High-Temperature Exposure of Photovoltaic Modules. 34th IEEE PVSC in Pennsylvania, USA (7–12 June 2009).

14 Kurtz, S., Whitfield, K., TamizhMani, G. et al. (2011). Evaluation of high-temperature exposure of photovoltaic module. *Progress in Photovoltaics* https://doi.org/10.1002/pip.1103.

15 Koehl, M. and Heck, M. (2009). Load evaluation of PV modules for outdoor weathering under extreme climatic conditions. *Proceedings of the 4th European Weathering Symposium*, Budapest, Hungary (16–18 September 2009).

16 Dixon, R.R. (1980). Thermal aging predictions from an Arrhenius plot with only one data point. *IEEE Transactions on Electrical Insulation* EI-15 (4): 331.

17 Kempe, M.D. (2006). Modeling of rates of moisture ingress into photovoltaic modules. *Solar Energy Materials and Solar Cells* 90: 2720–2738.

18 Kempe, M.D., Kurtz, S.R., Wohlgemuth, J. et al. (2011). Modeling of damp heat testing relative to outdoor exposure. PVSEC-21 in Fukuoka, Japan (30 November 2011).

19 Wohlgemuth, J.H. and Kempe, M.D. (2013). Equating Damp Heat Testing with Field Failures of PV Modules. 39th IEEE PVSC in Florida, USA (16–21 June 2013).

20 Kurtz, S., Wohlgemuth, J., Kempe, M. et al. (2013). Photovoltaic Module Qualification Plus Testing. NREL/TP-5200-60950.

21 Wohlgemuth, J. and Kurtz, S. (2014). Photovoltaic Module Qualification Plus Testing. 40th IEEE PVSC in Colorado, USA (8-13 June 2014).

22 Kato, K. (2012). PVRessQ! PV Module Failures Observed in the Field. NREL PVMRW.

23 Wohlgemuth, J.H., Cunningham, D.W., Amin, D. et al. (2008). Using Accelerated Tests and Field Data to Predict Module Reliability and Lifetime. 23rd EU PVSEC in Valencia, Spain (1–5 September 2008).

24 IEC 62892 (2019). Extended thermal cycling of PV modules – Test procedure.

25 IEC TS 62782 (2016). Photovoltaic (PV) modules – Cyclic (dynamic) mechanical load testing.

26 Wohlgemuth, J.H., Cunningham, D., Placer, N. et al. (2008). The Effect of Cell Thickness on Module Reliability. 33rd IEEE PVSC in California, USA (11–16 May 2008).

27 IEC TS 62804-1 (2015). Photovoltaic (PV) modules – Test methods for the detection of potential-induced degradation – Part 1: Crystalline silicon.

28 ASTM E2481-06 (2006). Standard Test Method for Hot Spot Protection Testing of Photovoltaic Modules.

29 IEC TS 62941 (2016). Terrestrial photovoltaic (PV) modules – Guideline for increased confidence in PV module design qualification and type approval.

30 CSA/ANSI C450-18 (2018). Photovoltaic (PV) module testing protocol for quality assurance programs.

31 IEC 63209 (to be published). Extended-stress testing of photovoltaic modules for risk analysis.

10

What does the Future Hold for PV and a Brief Summary

The discussions in the previous chapters looked at both the history and present-day status of photovoltaics (PV) module reliability and the accelerated stress testing associated with it. In this final chapter, the focus will shift to the future. PV is a dynamic industry so the technology and the testing standards are constantly evolving. The first section in this chapter will provide an update on the changes already in progress for some of the more important module qualification and safety standards. The second section will switch to a longer-range view, discussing how PV module reliability is likely to change in the future and what sort of accelerated stress testing will be necessary to validate the quality of the huge volume of modules produced. The book will end with a brief summary of the status of PV module reliability today.

10.1 Current Work on Updating Standards

Since PV is such a dynamic industry there is always work ongoing to update the standards. At this time, there are three important module testing standards in various stages of preparation. Two of these are updates of our old friends, the IEC 61215 series on Module Qualification and the IEC 61730 series on Module Safety. The third document is IEC TS 63126 [1] on qualifying modules to operate at higher temperatures. This is a new technical specification under development in WG2. Each of these will make significant changes to the way modules are tested.

10.1.1 Second Edition of IEC 61215 Series

Modifications to the 2016 version of the IEC 61215 series began as an effort to prepare an amendment, but the changes became so extensive that a new edition is required. The new edition will add several new tests, explain how several types of specialized modules are to be tested, clarify the details of the performance measurement pass/fail criteria, remove some characterization requirements and, as typical in a new edition, makes some clarifications. All six volumes of IEC 61215 (IEC 61215-1, IEC 61215-1-1, IEC 61215-1-2,

Photovoltaic Module Reliability, First Edition. John H. Wohlgemuth.
© 2020 John Wiley & Sons Ltd. Published 2020 by John Wiley & Sons Ltd.

IEC 61215-1-3, IEC 61215-1-4, and IEC 61215-2) require changes for this edition. Some of the proposed changes are discussed below.

- Potential Induced degradation (PID) (MQT 21) has been added as a new test leg requiring four modules. The test is conducted using the procedure given in IEC TS 62804-1 [2] but with chamber operating conditions of 85 °C, 85% RH. Two modules each are bias at plus and minus the maximum system voltage. The duration of the test is 96 hours with the typical pass/fail criteria including a maximum allowable power loss of 5%. This test is to be applied to all flat plate PV modules including cry-Si and thin films. It is hoped that adding this to the qualification test sequence will identify and eliminate from the market those modules susceptible to PID.
- Cyclic (Dynamic) Mechanical Load Testing (MQT 20) has been added into Sequence C between the UV Preconditioning (MQT 10) and the 50 thermal cycles (MQT 11). The test is conducted using the procedure given in IEC TS 62782 [3] with the recommended conditions of 1000 cycles with a 1000 Pa load. This test was added in an effort to identify modules that are susceptible to breakage especially of cry-Si cells.
- Special requirements necessary to conduct the test sequence on flexible modules have been added. This included the addition of a Bending Test (MQT 22) to ensure that a flexible module type can be bent without suffering permanent damage. Other changes for flexible modules include instructions on how to perform the mechanical tests and how to mount them during the chamber tests. For flexible modules, the document instructs the tester to mount the modules per the manufacturer's documentation with prescribed substrate and adhesive or attachment/mounting means during the test.
- Allows the use of representative samples when the modules under test are larger than the available simulators. Today, a module is considered very large if it exceeds 2.6 m in any dimension, or exceeds 2.1 m in both dimensions. In this case, a representative sample may be used for the chamber tests and Gate 2, the power measurements before and after stress testing. A full-sized module must still be measured for Gate 1 to validate the Standard Test Conditions (STC) power of the product. The representative samples should be as similar as possible to the full-size module in all electrical, mechanical, and thermal characteristics. The cell strings must be long enough to include bypass diodes and have 10 or more cells. The encapsulant, interconnects, electrical terminations, and the clearance distances around all edges should be the same as on the actual full-sized modules. There is an additional requirement that the representative samples can be no less than one-half the size of the full-sized module. If representative samples are used, the test report must include a table that lists the dimensions of the module type being qualified, as well as the dimensions of the samples tested for each of the stress tests (Module Quality Tests (MQTs)) for which a representative sample was used.
- Special requirements necessary to conduct the test sequences on bifacial modules, which are modules that can convert light received on both the front-side and rear-side into electricity, have been added. The modifications to IEC 61215 include:
  1) Procedures for measuring bifacial modules are provided as given in IEC TS 60904-1-2 [4]. Irradiance levels for the Gate 1 and Gate 2 Tests (See Section 4.3) were set at STC ($1000 \, W/m^2$) for the front side and $135 \, W/m^2$ for the rear-side.

2) The test levels have been increased during certain tests that require irradiance or current flow during the course of the test, because bifacial modules utilize some of the additional irradiance that is incident on the back side. A worst-case level of $300\,W/m^2$ was selected to set these test levels. The chosen levels are shown in Table 10.1. So, the irradiance for the hot spot test is increased to a level equivalent to STC ($1000\,W/m^2$) on the front plus $300\,W/m^2$ on the rear. For the other three tests listed (thermal cycling, bypass diode thermal test and bypass diode functionality test), the current level during the test is increased to the equivalent level produced with STC ($1000\,W/m^2$) on the front plus $300\,W/m^2$ on the rear.

3) Addition of a UV preconditioning test on the rear side. So Sequence C starts with the UV Preconditioning (MQT 09) on the front followed by the UV Preconditioning on the rear (MQT 09).

4) Additional reporting requirements for providing the bifaciality coefficients. Bifaciality coefficients are defined as the ratios between the I-V characteristics of the rear-side and the front-side of a bifacial module each measured under STCs. Bifaciality coefficients of short-circuit current, open-circuit voltage and peak power must be measured and reported.

5) The bifacial output power rather than the standard front side only measurement of power is used to determine pass/fail in Gates 1 and 2.

- Added a requirement that a 5 N weight be hung from the junction box during the thermal cycle test (MQT 11). This test was added because poor adhesion of junction boxes to modules has been observed in both fielded modules and during accelerated stress tests [5–7].
- Removed the requirement to measure the Nominal Module Operating Temperature (NMOT) and the requirement to measure the performance at NMOT. This was done for a number of reasons:

1) Measurement of NMOT is included in IEC 61853-2 [8].

2) The object of IEC 61215 is module qualification not characterization. The characterization tests were added into the early versions of IEC 61215 because they were not specified in any other standards. They are now so can be taken out of IEC 61215.

Table 10.1 Proposed current levels for testing in new edition of IEC 61215.

Stress Test	MQT #	Single Sided Module	Bifacial Module
Hot Spot Test	09	Irradiance $= 1000\,W/m^2$	Irradiance $=1300\,W/m^2$
Thermal Cycling	11	Applied Current $= I_{mp}$@STC	Applied Current $= I_{mp}$@ (STC front $+300\,W/m^2$ back)
Bypass Diode Thermal Test	18.1	Test Current $= 1.25 \times I_{sc}$ @ STC	Test Current $= 1.25 \times I_{sc}$ @ (STC front $+300\,W/m^2$ rear)
Bypass Diode Functionality Test	18.2	Test Current $= 1.25 \times I_{sc}$ @ STC	Test Current $= 1.25 \times I_{sc}$ @ (STC front $+300\,W/m^2$ rear)

Source: from 2019 draft of second edition of IEC 61215-1.

3) Accurate measurement of NMOT can take many months and delay the qualification of a new product.
4) Very few designers now use NMOT or the earlier NOCT. Typically, they have a thermal module that predicts module operating temperatures.

- The second edition will revise a portion of the monolithically-integrated (MLI) hot spot endurance test (MQT 09) to correct errors in the 2016 edition. The 2016 edition incorrectly mixed some of the procedures for testing wafer-based modules with testing of MLI modules. Minor revisions were also made in places where the procedure was unclear or may even have been impossible to perform.
- Sequence A (the characterization sequence) has been simplified by reducing the number of modules from three to one and explicitly stating that the measurements of sequence A may be performed in any order, or may be performed on two separate modules, if desired.
- Two major changes have been proposed for the Insulation Test (MQT 03). The first change involves the procedure. The 2016 version made a change that said "If the module has no frame or if the frame is a poor electrical conductor, wrap a conductive foil around the edges. Cover all polymeric surfaces (frontsheet, backsheet and junction box) of the module with conductive foil." There has been concern with this procedure as it is hard to detect surface tracking when all polymeric surfaces are covered with foil. It is also likely that the capacitance introduced by covering a large area with foil disables the test equipment's arc discharge detection. So the new edition proposes to return to the 2005 version of the test, but removes the 2005 requirement of covering the entire back of a frameless module with foil. The second change in the Insulation test is to set the test levels in harmony with IEC 61730. So the test levels are now based on the Module Class and whether the module uses a cemented joint. Table 10.2 shows how the test levels will be set for MQT 03.

Where:

V_{Test1} = one-minute high voltage stress
V_{Test2} = two-minute applied voltage for measuring Insulation Resistance
V_{sys} = Manufacturer's maximum rated system voltage for the module type being tested

Module Classes are from IEC 61730.

For module types that are certified to IEC 61730 using cemented joints the one-minute high voltage stress levels must be increased to 135% of the values given in Table 10.2.

Table 10.2 Voltage Stress Levels for insulation Test (MQT 03).

Module Class	V_{Test1} **(volts)**	V_{Test2} **(volts)**
0	$1000 + 2 \times V_{sys}$	Greater of 500 or V_{sys}
II	$2000 + 4 \times V_{sys}$	Greater of 500 or V_{sys}
III	500	500

Source: from 2019 draft of second edition 2 of IEC 61215–2.

- The power ratings and power measurement pass/fail criteria for Gate 1 have been clarified in the new edition. Changes have been proposed in the following two areas.

  1) Gate No. 1 criteria require use of the module manufacturer's rated values and tolerances from the nameplate and product datasheet. Rated values and tolerances for module power, short-circuit current, and open-circuit voltage are all required. There has been confusion as to how the different values have been presented by different module manufacturers. To clarify this, the draft contains examples of module labels and then interprets them, providing the nominal values and tolerances for power, short-circuit current and open-circuit voltage.

  2) Gate 1 is used to verify that the module type meets its rated output power. In the first edition of IEC 61215-1, all of the modules measured had to meet the minimum value specified for the product. The second edition recognizes that there are statistical variations in module performance and measurements. If less than 10 modules are measured, they all have to pass the Gate 1 requirements for individual module measurements. However, if 10 or more modules are measured, a single module failure, that is one measurement below the minimum value is allowed.

- Changes have been proposed to the requirements for the simulator used for making maximum power measurements. The first edition requires either use of a Class BBA or better solar simulator according to IEC 60904-9 [9] with a matched reference device, which isn't always available or a choice of making the spectral correction calculation according to IEC 60904-7 [10] with the BBA simulator or using a simulator rated AAA according to IEC 60904-9. The problem with this requirement is that using an AAA simulator doesn't guarantee a particular accuracy and it is not always possible to make the spectral correction since you may not know the spectral response of the test modules. In the new edition, the following requirements have been proposed for the simulator:

  1) Type A spatial uniformity is required to reduce measurement errors, but the spectral classification can be as low as a C, as long as the requirements on uncertainty given in 3 below are met.

  2) There are now multiple, acceptable ways to obtain the spectral response data for the mismatch correction. Spectral response data may be taken by any test lab that is accredited for that measurement. The device used to obtain the spectral response data can be either one of the test modules or it may be a packaged cell made with the same components as the test modules. The spectral response measurement is performed in accordance with IEC 60904-8 [11].

  3) The component of uncertainty due to spectral mismatch must be included in the total measurement uncertainty used in evaluating Gate No. 1. Maximum allowable values for the total uncertainty are now provided in the technology-specific parts (3% for cry- Si, 4% for single junction thin films and 5% for multi-junction thin films).

A final reminder that all of these changes are still in process, so the final published the second edition of the IEC 61215 series may be somewhat different from this summary of the draft.

10.1.2   Amendment 1 to Second Edition of IEC 61730-1 and IEC 61730-1

At the time that second edition of IEC 61730 was published, there were several test procedures that the industry wished to include but had not been finalized. Rather than hold up the document, the second edition was published with placeholders. Since the publication, a project team has been working on an amendment to update the module safety standard. The proposed additions to the standard are discussed below:

- Requirements that specific components like the junction boxes, connectors, backsheets and frontsheets be qualified according to their component standards.
- Replacement of weathering tests for frontsheets/backsheets with a requirement that relied-upon insulation be qualified according to IEC 62788-2-1 [12]
- Removal of all reference to "open-rack" mounting configuration, being replaced by the 98th percentile application temperature as defined in the new High Temperature Standard (See next subsection).
- Clarification of what Distance through Insulation (DTI) means and how it is measured.
- Modules designed for Class 0 that are for use in restricted access areas would no longer be required to pass the module breakage test (MST 32).
- Substitution of a different procedure for the Sharpness of Edges Test (MST 06).
- Clarification of the procedure for the Ignitability Test (MST 24)
- Corrections and clarifications in many areas especially for the test descriptions in IEC 61730-2.

Once again, a reminder that all of these changes are still in process so the final published Amendment of Edition 2 of IEC 61730-1 and IEC 61730-2 may differ from the summary of the draft given here.

10.1.3   IEC TS 63126 – Guidelines for Qualifying PV Modules, Components and Materials for Operation at High Temperatures

The IEC qualification and safety standards (IEC 61215 series and IEC 61730 series) provide testing requirements for PV modules suitable for long-term operation in general open-air climates as defined in the first edition of IEC 60721-2-1 [13]. General open-air climates have an absolute maximum temperature of +45 °C. There are places in the world (think Death Valley, California and the Saudi Arabian peninsula) where this limit is often exceeded. We also know that modules mounted in locations with restricted cooling, like rooftops or building integrated; have higher operating temperatures than for open-rack mounting with unrestricted cooling. Work at Arizona State University showed that roof-mounted modules often reach in excess of 90 °C [14]. Work led by Michael Kempe of NREL demonstrated module temperatures in excess of 100 °C for modules deployed in Arizona with insulated backs [15]. So the testing in IEC 61215 and 61730 and the testing in some of standards for components used in PV modules cover only a portion of the temperature range that PV modules are likely to encounter during outdoor operation. Therefore, TC82 set out to write a guideline to specify what additional testing may be required to qualify PV modules for operation at higher temperatures.

IEC TS 63126 is designed to provide guidance on how to modify the testing in IEC 61215, IEC 61730, IEC 62790 [16] and IEC 62852 [17] to account for the fact that the PV modules may operate at temperatures in excess of those assumed in the referenced standards. The project team reviewed all of the stress tests in these documents and decided whether the stress level had to be increased and if so how.

One of the main questions for the project team was how to define the module operating temperature categories. Do you base the testing temperature level on the maximum temperature ever observed or some yearly average of the daily maximum? The problem with using the absolute maximum is that the module is only at this temperature for a few minutes and so not much degradation can occur. The team under Kent Whitfield from UL, decide to use the concept of 98% temperature. For any given application, defined by a geographic location and a mounting system, the module would be at or below the 98% temperature 98% of the time. Or put another way, the module's temperature would exceed the 98% temperature only 2% of the time or for 175 hours per year.

A second question is how many different categories or levels do we want to test for? It would not be practical to have a separate set of qualification and safety tests for every high temperature application. As an arbitrary answer to this the group selected three categories:

1) Open-rack mount systems in general open-air climates: This is the temperature range already covered by the standards and has a 98th percentile temperature of less than 70 °C (See for example Figure 9.1) [18, 19].
2) Roof-mount systems and open-rack systems in higher temperature locations: This is defined as Temperature Level 1 with 98th percentile temperature in a range between 70 and 80 °C.
3) Insulated back modules: This is defined as Temperature Level 2 with 98th percentile temperature in a range between 80 and 90 °C.

These temperature categories have then been used to modify the testing requirements.

The proposed changes to IEC 61215 for higher temperature operation are detailed in Table 10.3. As you can see, in most cases, the temperature of the test was increased by 10 °C for Level 1 and 20 °C for Level 2. The only exception was the bypass diode test where the work by the Photovoltaic Quality Assurance Task Force (PVQAT) Task Group 4 provided inputs to select the revised test temperatures and current flows [20, 21]. The project team decided that the humidity-based tests, the Damp Heat (MQT 13) and Humidity-Freeze (MQT 12) tests, should not be changed. Higher module temperatures usually dry the modules

Table 10.3 Proposed changes to IEC 61215 to take higher operating temperatures into account.

Test	MQT #	Original	Level 1	Level 2
Hot Spot	09	50 ± 10 °C	60 ± 10 °C	70 ± 10 °C
UV Preconditioning	10	60 ± 5 °C	70 ± 5 °C	80 ± 5 °C
Thermal Cycling – upper limit	11	85 ± 2 °C	95 ± 2 °C	105 ± 2 °C
Bypass Diode Thermal –chamber temperature	18.1	75 ± 2 °C	90 ± 2 °C	100 ± 2 °C
Part 1 – Current		I_{sc}	$1.15 \times I_{sc}$	$1.15 \times I_{sc}$
Part 2 – Current		$1.25 \times I_{sc}$	$1.4 \times I_{sc}$	$1.4 \times I_{sc}$

Table 10.4 Proposed changes to IEC 61730 to take higher operating temperatures into account.

Test	MST #	Original	Level 1	Level 2
RTI/RTE/TI		≥90 °C	≥100 °C	≥110 °C
Hot Spot	22	50±10 °C	60±10 °C	70±10 °C
UV	54	60±5 °C	70±5 °C	80±5 °C
Thermal Cycling – upper limit	51	85±2 °C	95±2 °C	105±2 °C
Material Creep	37	105 °C	105 °C	110 °C
Dry Heat Conditioning	56	105 °C	105 °C	110 °C

out so the Level 1 and 2 humidity stresses should be no worse than for the open-rack mount modules in the original standard.

The proposed changes to IEC 61730 for higher temperature operation are detailed in Table 10.4. Once again, in most cases, the temperature of the test was increased by 10 °C for Level 1 and 20 °C for Level 2. The only exceptions being the last two, namely the Material Creep Test and Dry Heat Conditioning. In both of these cases, the original test in the 2016 version of IEC 61730 was performed at a temperature of 105 °C. Certainly the open-rack, general open-air climate modules would never exceed this nor would the Level 1 modules. However, Level 2 modules could so the material creep and dry conditioning temperatures for this level were increased to 110 °C since this appears to be the maximum temperature that Level 2 modules are likely to experience for any extended time period.

The situation with the two component standards (IEC 62790 for junction boxes and IEC 62852 for PV connector) is somewhat different in that they specify the test levels based on the ratings that the manufacturer puts on the product. For IEC 62790, this value is called the Upper Ambient Temperature. For PV connectors, this value is called Upper Limit Temperature (ULT). There are no requirements in the standards for any minimum (or maximum) values for these two temperature ratings. In the new high temperature guidelines, it is being proposed that the Upper Ambient Temperature and the ULT be a minimum of 95 °C for Level 1 modules and a minimum of 105 °C for Level 2 modules. These values will then define the temperatures at which the stress tests in the two standards are performed.

The final area under review for IEC TS 63126 is the evaluation of at what temperatures long-term UV testing of the materials for use in PV modules should be performed. The actual requirements for the different materials (backsheets, encapsulants, etc.) have yet to be determined. The UV exposure levels should be chosen from IEC 62788-7-2 [22]. Table 10.5 contains a set of recommended UV exposure temperatures taken from IEC 62788-7-2 [22]. As you can see, each higher class is 10 °C hotter than the previous one. It has been proposed that the high temperature guidelines just tell the user to use one category higher for Level 1 modules and two categories higher for Level 2 modules. So if the final specification says that backsheets should be tested using A3, then Level 1 modules should be tested using A4 and Level 2 modules tested using A5.

This higher temperature of operation guideline IEC TS 63126 is scheduled for publication in 2020. After the industry has had a few years to work with these requirements, it is

Table 10.5 Temperatures of UV exposure conditions from IEC 62788-7-2.

Condition #	Chamber Air Temperature	Black Panel Temperature
A1	45 °C	70 °C
A2	55 °C	80 °C
A3	65 °C	90 °C
A4	75 °C	100 °C
A5	85 °C	110 °C

expected that the higher temperature testing requirements will be incorporated directly into the relevant qualification or safety standards. Adoption of these guidelines should help to eliminate some of the premature failures that have been observed in modules operating at higher temperatures.

10.2 Looking to the Future

The PV market has continued to grow with several estimates indicating that 2018 world-wide module shipments exceeded 100 Gigawatts [23, 24]. There is room in the future for continued growth, but not the many orders of magnitude growth experienced in the past. As PV begins to provide a significant percentage of the world's overall supply of electricity, production volumes will likely level off. After this, the level of production will stabilize based on meeting the increasing worldwide demand for electricity plus the replacement of modules from old PV systems and from some retired fossil fuel plants. This level is likely to be greater than today's 100 Gigawatts of annual production, but probably not by orders of magnitude. Based on a future volume scenario of about an order of magnitude larger than today's annual production, let's consider what the PV reliability programs should look like.

Reliability is being addressed as one of the ways to reduce the cost of PV supplied electricity. One of the factors in this analysis is the annual degradation of PV modules and PV power plants. The second way reliability impacts cost of PV electricity is through the module lifetime. Let's take a look at these two factors.

10.2.1 Degradation Rates

Today, a degradation rate of 0.5% per year is typically used in the PV cost models. This may be aggressive as Jordan [25] has shown that while the mean degradation rate of the systems he has studied is 0.5%, the average PV systems degradation rate is 0.8% and many systems are worse (as shown in Figure 6.1). In reality, system power degradation is a combination of module power degradation, premature module failures and other system degradations. So, for modules we are worried about both the annual degradation rate and any premature failures.

Qualification Testing and use of Quality Management Systems (QMS) help reduce/eliminate the premature failures. Therefore, it is critical that the Qualification Tests be regularly updated to take new failure modes and new technologies into account. As our previous section indicated, IEC 61215 is now being updated to take PID, cell breakage, bifacial modules and flexible modules into account. In the past few years, bifacial modules have become an important technology for use in large power systems so getting this technology included in the Qualification Test Sequence is important. As other changes occur, for example, the switch from using front encapsulants with UV blockers to front encapsulants without the UV blockers, it is important to evaluate the possible impact on reliability beyond the qualification level. For example, in the encapsulant case, we should be determining whether without the UV blockers, UV will degrade the adhesion at the cell surface. As new failure modes are identified from the field, new accelerated stress test sequences must be developed to duplicate them. Once established some of these new tests should be incorporated into the Qualification Tests to eliminate those failures from future generations of modules.

The degradation rates on the other hand are still somewhat of an unknown. This is one area that needs work in the near future. We need to better define why PV modules are degrading in the field and to develop accelerated stress tests that duplicate those degradation processes and rates. These new accelerated stress tests should then be added to our set of extended stress tests that was discussed in Chapter 9.

10.2.2 Module Lifetime

The second reliability factor that impacts the cost of PV electricity is the module lifetime. This actually has less impact than the degradation rate in the cost model because net present value analysis means income collected far in the future is not worth as much as money earned in the near future. On the other hand, in practical terms, if you can extend the life of a PV array from 25 to 30 years you can expect to collect 5 more years of electricity from it (minus the loss to degradation). Five years of electricity has significant monetary value. So better understanding of the module lifetime and developing a methodology to measure it as discussed in Chapter 9 is important. For module manufacturers, this implies not only the effort to establish a program to actually measure the module lifetime but also a degree of control and stability of the design and manufacturing processes. That means not always chasing the latest way to save a little cost but rather finding a stable and cost-effective design and manufacturing system and sticking with it for an extended period of time. Modifications can be bundled and then tested extensively before being implemented into volume production.

The last thing to discuss is the impact of new PV technologies. In 2018, more than 90% of the PV modules manufactured and shipped around the world were cry-Si. So newer technologies, including thin films that have been around now for decades are still struggling to gain market share. The present generation of commercial thin-film technologies (CdTe, CIGS, and a-Si) have decades of field experience and now pass the same IEC 61215 Qualification Tests as cry-Si module. However, there are still some concerns about the long-term reliability and lifetimes of these products. For example, will the edge seals survive and keep the moisture out of the module for 25 years? A set of accelerated stress tests are now under development at NREL to evaluate the durability of the edge seals, which is leading to development of IEC 62788-5-2 [26]. Use of this test sequence, combined with continued field experience, should increase confidence in this generation of thin-film products.

What about the newer technologies – organics, dye-sensitized and perovskites? These all appear to offer very low-cost solutions to PV, but they all have issues with instability. Research is still struggling to find materials and device structures that are stable enough to be repeatedly measured. Until such stable devices are developed, it is really too early to discuss reliability or lifetime. Once a significant degree of stability has been reached, this new generation of materials must follow the same path as earlier generations of PV. So devices must be fielded, failure modes identified, new accelerated stress tests developed and then used to improve the product reliability and extend the lifetime.

10.3 Brief Summary

As was stated in Chapter 1 of this book, the perceived reliability and long expected life of modules has been critical to PV's commercial success. Investors are not likely to risk billions of dollars to purchase PV modules if they don't believe that those modules will survive and produce the expected amount of electricity. The major aim of this book was to explain how this reliability and extremely long lifetime came about. It may have been important that PV markets grew slowly in the early days so that the feedback loop between field failures, accelerated stress tests and improved products could occur during periods of low volume production. By the time PV systems grew to the Megawatt scale, the Qualification Tests were in use, module warranties were greater than 10 years and modules generally performed as specified.

In the last 10 years, the dramatic increase in module production volume in Asia, coupled with the corresponding price reductions have revolutionized the PV industry. Now we are seeing hundred Megawatt systems installed around the world with annual production volumes now in excess of 100 Gigawatts. There are even claims that PV is now the least expensive form of electricity that can be installed [27]. The rapid growth in production volume is amazing enough, but the real miracle is that this increase was achieved without appreciably impacting the reliability of the products. This can be at least partially attributed to the mindset in PV that modules must be qualified to IEC 61215 and IEC 61730. The new module manufacturers understood that they needed to have their modules certified to IEC 61215. Tamizhmani reported that the failure rates in qualification testing increased significantly in the time period from 2005 to 2007 when many of the new Chinese module manufacturers were setting up operation [28]. Most of these manufacturers worked through their issues and received certification for their modules. Qualification certification was so important that several counterfeit IEC 61215 certificates were identified on line. Thankfully, by designing and building modules to meet IEC 61215, these new suppliers had products that also performed well in the field.

One of the final messages of the book is that while PV modules are extremely reliable, PV reliability work is not finished. As PV continues to expand, providing an ever-larger percentage of the world's electricity, the reliability and standards communities must continue to support these efforts through:

- Continued updating of the Qualification Tests using the latest results on field performance and for new technologies.
- Improved understanding of why modules degrade in the field providing important inputs into selection of one accepted set of extended stress tests.
- Development and implementation of a methodology for determining module service life in specific geographic locations.

These efforts should continue as long as commercial PV is continuing to modify their processes and materials, which is likely to be forever.

10.3.1 Personal Reflections

Having joined Solarex in 1976, in the early days of PV, I have been asked on numerous occasions "when did I realize that PV had arrived?" In looking back on this more than 40-year period, I can remember a lot of "wins" for PV. I always thought it would be when all of our world leaders recognized that PV was a viable, low-cost source of electricity. But this doesn't appear to have happened yet as, within the last year, I heard one of our "Well informed" Congressmen explain that PV was still not cost effective or reliable.

Maybe PV had arrived when people stopped asking me "where did the hot water come out of the panel?" Or maybe it was when my new word processing software actually contained the word "photovoltaics" in its spell-checking dictionary rather than to automatically correct it to "prophylactics" like my earlier version did. Or, maybe it was in 1996, when Siemens Solar handed out pins proclaiming that they had shipped a cumulative total of 100 Megawatts of PV modules. Or maybe it was when annual shipments topped a Gigawatt or maybe 100 Gigawatts. I think the best answer is that it occurred when we began to see reports that PV electricity was cost competitive with building conventional fossil fuel plants. This is really what the industry had been working toward for more than a generation and its achievement is a credit to all those who have contributed.

References

1 IEC TS 63126 (Publication expected in 2020). Guidelines for qualifying PV modules, components and materials for operation at high temperatures.
2 IEC TS 62804-1 (2015). Photovoltaic (PV) modules – Test methods for the detection of potential-induced degradation – Part 1: Crystalline silicon.
3 IEC TS 62782 (2016). Photovoltaic (PV) modules – Cyclic (dynamic) mechanical load testing.
4 IEC TS 60904-1-2 (2019). Photovoltaic devices – Part 1–2: Measurement of current-voltage characteristics of bifacial photovoltaic (PV) devices.
5 Gambogi, W., Kurian, S., Hamzavytehrany, B. et al. (2011). The Role of Backsheet in Photovoltaic Module Performance and Durability. EU PVSEC in Hamberg, Germany (5–9 September 2011).

6 Sample, T., Skoczek, A., and Field, M.T. (2009). Assessment of Ageing Through Periodic Exposure to Damp Heat (85°/85% RH) of Seven Different Thin-Film Module Types. 34th IEEE PVSC in Pennsylvania, USA (7–12 June 2009).

7 Miller, D.C. and Wohlgemuth, J.H. (2012). Examination of a Junction-Box Adhesion Test for Use in Photovoltaic Module Qualification. SPIE Optics + Photonics in California, USA (12–16 August 2012).

8 IEC 61853-2 (2016). Photovoltaic (PV) module performance testing and energy rating – Part 2: Spectral responsivity, incidence angle and module operating temperature measurements.

9 IEC 60904-9: Edition 2 (2007). Photovoltaic devices – Part 9: Solar simulator performance requirements.

10 IEC 60904-7: Edition 3 (2008). Photovoltaic devices – Part 7: Computation of the spectral mismatch correction for measurements of photovoltaic devices.

11 IEC 60904-8: Edition 3 (2014). Photovoltaic devices – Part 8: Measurement of spectral responsivity of a photovoltaic (PV) device.

12 IEC 62788-2-1 (Publication expected in 2021). Polymeric materials for photovoltaic (PV) modules – Part 2–1: Safety requirements for polymeric frontsheet and backsheet.

13 IEC 60721-2-1: Edition 1 (2002). Classification of environmental conditions – Part 2–1: Environmental conditions appearing in nature – Temperature and humidity.

14 Oh, J., Tamizhmani, G., and Palomino, E. (2010). Temperatures of building applied photovoltaic (BAPV) modules: Air gap effects. *Proceedings of SPIE – The International Society for Optical Engineering*, California, USA (1–5 August 2010).

15 Kempe, M.D., Miller, D.C., Wohlgemuth, J.H. et al. (2015). Field testing of thermoplastic encapsulants in high temperature installations. *Energy Science & Engineering* 3 (6): 565–580.

16 IEC 62790 (2014). Junction boxes for photovoltaic modules – Safety requirements and tests.

17 IEC 62852 (2014). Connectors for DC-application in photovoltaic systems – Safety requirements and tests.

18 Kurtz, S., Whitfield, K., Miller, D. et al. (2009). Evaluation of Cumulative High-Temperature Exposure of Photovoltaic Modules. 34th IEEE PVSC in Pennsylvania, USA (7–12 June 2009).

19 Kurtz, S., Whitfield, K., TamizhMani, G. et al. (2011). Evaluation of High-Temperature Exposure of Photovoltaic Module. *Progress in Photovoltaics* 19 (8): 954–965.

20 Shiradkar, N.S., Gade, V.S., Schneller E.J., and Sundaram, K.B. (2015). Revising the Bypass Diode Thermal Test in IEC 61215 Standard to Accommodate Effects of Climatic Conditions and Module Mounting Configurations. 42nd IEEE PVSC in Florida, USA (14–19 June 2015).

21 Shiradkar, N.S., Gade, V.S., Schneller, E.J., and Sundaram, K.B. (2018). Revising the Bypass Diode Test to Incorporate the Effects of Photovoltaic Module Mounting Configuration and Climate of Deployment. WCPEC-7 in Hawaii, USA (10–15 June 2018).

22 IEC 62788-7-2 (2017). Measurement procedures for materials used in photovoltaic modules – Part 7–2: Environmental exposures – Accelerated weathering tests of polymeric materials.

23 Fraunhofer (2019). FraunhoferInstitute for Solar Energy Systems, ISE Photovoltaics Report – 2018. https://www.ise.fraunhofer.de/content/dam/ise/de/documents/publications/studies/Photovoltaics-Report.pdf (Accessed 29 August 2019).

24 Bloomberg (2019). FraunhoferInstitute for Solar Energy Systems, ISE Photovoltaics Report – 2018. https://about.bnef.com/blog/clean-energy-investment-exceeded-300-billion-2018 (Accessed 29 August 29, 2019).

25 Jordan, D.C. and Kurtz, S.R. (2013). Photovoltaic Degradation Rates — An Analytical Review. *Progress in Photovolatics Research and Applications* 21 (1): 12–29.

26 IEC 62788-5-2 (Publication expected in 2020). Measurement procedures for materials used in photovoltaic modules – Part 5–2: Edge seals – Edge-seal durability evaluation guideline.

27 Popular Mechanics (2016). https://www.popularmechanics.com/science/green-tech/a24357/solar-power-cheapest-energy (Accessed 29 August 2019).

28 TamizhMani, M.G. (2010). Experience with Qualification and Safety Testing of Photovoltaic Modules. NREL PVMRW. https://www.nrel.gov/docs/fy14osti/60171.pdf (Accessed 29 August 2019).

Index

Photovoltaic Module Reliability, First Edition. John H. Wohlgemuth.
© 2020 John Wiley & Sons Ltd. Published 2020 by John Wiley & Sons Ltd.